AF880278

Lecture Notes in Mechanical Engineering

Lecture Notes in Mechanical Engineering (LNME) publishes the latest developments in Mechanical Engineering—quickly, informally and with high quality. Original research or contributions reported in proceedings and post-proceedings represents the core of LNME. Volumes published in LNME embrace all aspects, subfields and new challenges of mechanical engineering.

To submit a proposal or request further information, please contact the Springer Editor of your location:

Europe, USA, Africa: Leontina Di Cecco at Leontina.dicecco@springer.com
China: Ella Zhang at ella.zhang@springer.com
India: Priya Vyas at priya.vyas@springer.com
Rest of Asia, Australia, New Zealand: Swati Meherishi at swati.meherishi@springer.com

Topics in the series include:

- Engineering Design
- Machinery and Machine Elements
- Mechanical Structures and Stress Analysis
- Automotive Engineering
- Engine Technology
- Aerospace Technology and Astronautics
- Nanotechnology and Microengineering
- Control, Robotics, Mechatronics
- MEMS
- Theoretical and Applied Mechanics
- Dynamical Systems, Control
- Fluid Mechanics
- Engineering Thermodynamics, Heat and Mass Transfer
- Manufacturing Engineering and Smart Manufacturing
- Precision Engineering, Instrumentation, Measurement
- Materials Engineering
- Tribology and Surface Technology

Indexed by SCOPUS, EI Compendex, and INSPEC.

All books published in the series are evaluated by Web of Science for the Conference Proceedings Citation Index (CPCI).

To submit a proposal for a monograph, please check our Springer Tracts in Mechanical Engineering at https://link.springer.com/bookseries/11693.

Anish Sachdeva · Kapil Kumar Goyal ·
Rajiv Kumar Garg · J. Paulo Davim
Editors

Recent Advances in Operations Management and Optimization

Select Proceedings of CPIE 2023

Springer

Editors
Anish Sachdeva
Department of Industrial and Production Engineering
Dr. B. R. Ambedkar National Institute of Technology Jalandhar
Jalandhar, Punjab, India

Kapil Kumar Goyal
Department of Industrial and Production Engineering
Dr. B. R. Ambedkar National Institute of Technology Jalandhar
Jalandhar, Punjab, India

Rajiv Kumar Garg
Department of Industrial and Production Engineering
Dr. B. R. Ambedkar National Institute of Technology Jalandhar
Jalandhar, Punjab, India

J. Paulo Davim
Department of Mechanical Engineering
University of Aveiro
Aveiro, Portugal

ISSN 2195-4356 ISSN 2195-4364 (electronic)
Lecture Notes in Mechanical Engineering
ISBN 978-981-99-7444-3 ISBN 978-981-99-7445-0 (eBook)
https://doi.org/10.1007/978-981-99-7445-0

This Springer imprint is published by the registered company Springer Nature Singapore Pte Ltd.
The registered company address is: 152 Beach Road, #21-01/04 Gateway East, Singapore 189721, Singapore

Contents

About the Editors

Dr. Anish Sachdeva is a Professor in the Industrial and Production Engineering Department at Dr. B. R. Ambedkar National Institute of Technology Jalandhar, India. His areas of interest are reliability and maintenance engineering, supply chain management, quality management, human factor engineering, and simulation of production systems. He did his Ph.D. from the Indian Institute of Technology Roorkee (India). He is Lean Six Sigma Black Belt Expert certified by CII-IQ. He has over 28 years of industry and teaching experience. He has guided 65 M. Tech. and 17 Ph.D. students, presently supervising 06 Ph.D. scholars. He has about 200 publications in international and national journals and proceedings of international conferences to his credit. At present, he is a reviewer of many international journals. He is the guest editor of various journals published by Emerald Publishes, Elsevier and Inderscience publications. He has been designated as member of international and national advisory committees of various international and national conferences organized by various international and national organizations/institutions. He is associated with various engineering colleges in the region as a member of their advisory board/governing council. He has been member of the board of studies of various prestigious Institutions like NIT Delhi, PEC Chandigarh, NIT Kurukshetra, PTU Kaputhala, etc. He has organized six international conferences at NIT Jalandhar as organizing secretary and three international conferences as Chairman and holds the position of Director (Honorary), Quality Circle Forum of India. He has visited the USA, the UK, Norway, Malaysia, Thailand, Hong Kong, France and Mauritius and delivered his research papers at various international conferences. At NIT Jalandhar, he has handled various portfolios like Dean Academic, Dean Student Welfare, Public Relation Officer, Chief Warden, Nodal Officer TEQIP-II/III, etc. from time to time during the past 28 years of his service.

Dr. Kapil Kumar Goyal is serving as an Assistant Professor in the Department of Industrial and Production Engineering at Dr. B. R. Ambedkar National Institute of Technology Jalandhar. He earned his bachelor's degree in Mechanical Engineering from NIT Kurukshetra and completed his master's degree in Manufacturing Technology from NITTTR Chandigarh, India. He obtained his doctorate from IIT Roorkee in 2012. Currently, with 20 years of teaching experience and 4 years of experience in the public sector and industry, he has established himself as a seasoned professional. Dr. Goyal has an impressive track record, having authored more than 40 publications that have been published by renowned publishers including Elsevier, Taylor & Francis, Springer, and Inderscience. He has received accolades for his research, including the best paper awards at international conferences. His research interests encompass a broad range of subjects, including evolutionary algorithms, multiple objective optimization, decision sciences, machine learning, manufacturing systems, additive manufacturing, industrial internet of things and process planning and optimization. He also serves as an Editor for various international journals and acts as Reviewer for several reputable publishers. In addition to his academic pursuits, he has delivered expert lectures at prestigious institutions like IITs and NITs.

Dr. Rajiv Kumar Garg is Professor in the Industrial and Production Engineering Department at Dr. B. R. Ambedkar National Institute of Technology Jalandhar, India. His areas of interest are non-traditional machining, supply chain management, industrial management, and systems dynamics. He holds a B.Sc. Engg. (Hons.) in Mechanical Engineering from Regional Engineering College Kurukshetra, an M.E. (Hons.) in Industrial Engineering, and a Ph.D. from Thapar Institute of Engineering and Technology Patiala (India). He has over 35 years of industry and teaching experience. He has guided 47 M.Tech. and 12 Ph.D. candidates to pursue their Ph.D. degrees. He has over 138 publications in international and national journals and proceedings of international conferences to his credit. He has been Reviewer of many international journals, Associate Editor of the Global Journal of Flexible Systems Management (Springer), and also has been the Guest Editor of the International Journal of Machining and Machinability of Materials (IJMMM). He has organized six international conferences at NIT Jalandhar as Chairman and Convener.

Dr. J. Paulo Davim is Full Professor at the University of Aveiro, Portugal. He is also distinguished as Honorary Professor in several universities/colleges/institutes in China, India, and Spain. He received his Ph.D. in Mechanical Engineering in 1997, M.Sc. in Mechanical Engineering (materials and manufacturing processes) in 1991, Mechanical Engineering (5 years) in 1986, from the University of Porto (FEUP), the Aggregate title (Full Habilitation) from the University of Coimbra in 2005 and the D.Sc. (Higher Doctorate) from London Metropolitan University in 2013. He is a Senior Chartered Engineer by the Portuguese Institution of Engineers with an MBA and Specialist titles in Engineering and Industrial Management as well as in Metrology. He is also Eur Ing by FEANI-Brussels and Fellow (FIET) of IET-London. He is Editor-in-Chief of several international journals, Guest Editor of journals, Books Editor, Book Series Editor, and Scientific Advisory for many international journals and conferences.

Design and Analysis of an Adaptive Robotic Gripper

Yashraj M. Patil, N. I. Jamadar, Lalit N. Patil, and Digvijay G. Bhosale

Abstract In this study, an inexpensive fundamental robotic end effector module was investigated in order to make it easier to conduct robotic grasping research. The three-finger under actuated robotic gripper was designed in three dimensions and produced using a servomotor actuator that is readily available off the market and a small number of 3D printed parts. Following a comprehensive explanation of an under actuated finger, gear train mechanisms, in addition to general gripper assembly design, an illustration and explanation of the grabbing of objects with varied geometries follow. For research and instructional reasons, robotic researchers can utilize the offered open-source gripper design as a base to create custom robotic end effectors. An innovative prototype and a thorough introduction as well as an explanation of the gripper design idea are given. Using a four-bar linkage mechanical system and a single off-the-shelf actuator, it is demonstrated that the proposed robotic gripper with a single actuator satisfies the proposed objectives for its simple mechanical structure, low cost, and relatively high payload compared to similarly sized tendon-driven robotic end effectors.

1 Introduction

Indian cities are becoming flatteringly more lively on a global scale, particularly in three crucial sectors: aerospace, defense, transportation, and support services [1–3]. An interdisciplinary discipline that combines science, engineering, and technology is robotics. It entails the creation, assembly, control, and employment of robots to accomplish jobs that have historically been completed by humans or to stand in

Y. M. Patil (✉) · N. I. Jamadar
Department of Mechanical Engineering, Dr. D. Y. Patil Institute of Technology, Pimpri, Pune 411018, India
e-mail: Yashraj.Patil@dypvp.edu.in

L. N. Patil · D. G. Bhosale
Department of Automation and Robotics, Dr. D. Y. Patil Institute of Technology, Pimpri, Pune 411018, India

A. Sachdeva et al. (eds.), *Recent Advances in Operations Management and Optimization*, Lecture Notes in Mechanical Engineering,
https://doi.org/10.1007/978-981-99-7445-0_1

for their actions. Robotics aims to create devices that can aid and support people [4]. It combines, among other disciplines, the subjects of computer science, control engineering, software engineering, mechatronics, electronics, bioengineering, and electrical and mechanical engineering [5, 6]. Robots are frequently utilized in fields like manufacturing automobiles to carry out basic, repetitive activities and in fields where work must be done in dangerous conditions for people [7–9].

The gripper has been an essential component ever since George DeVol created the first industrial robot in 1959. The concept of a non-human machine that can carry out duties that are typically performed by people is the foundation of the entire robot concept [10]. Applications for robotic grasping are numerous, including automated manufacturing lines in the industry and everyday support for the crippled [11, 12]. However, handling items of various sizes, shapes, and characteristics is a challenge for classic stiff grippers in many applications. Rigid grippers are often employed to properly determine the location and geometry of those items, which significantly raises the use-cost and restricts their use. Grasping tools engage with the work item, allowing communication between robotic systems including complexes as well as the environment in which they operate [4]. Gripper mechanisms transform input motion and force into the requisite output motion and gripping force as end-effectors of various technological devices [9]. Soft grippers have softer materials and inherently more flexible mechanical joints and linkages than rigid grippers do. Because of their flexibility and delicacy, they may securely interact with people and conform to the interface of the target item. The major focus of recent research has been on soft grasping, which may be actuated in a variety of methods, such as passive frameworks with auxiliary motors, shape materials [13]. Robotic systems have a significant problem when it comes to delicately gripping and manipulating things in dense, unstructured settings like those seen in agriculture, food processing, or the household. Robotic hands have difficulty gripping objects because of unknown contact types, object stiffness qualities, and modelling uncertainty for both the hand and the item [9, 14].

A gripper is indeed an end-of-arm tool attached to the machinery used to pick up, transport, and position workpieces. The gripper grabs and releases workpieces by opening and shutting its fingers using electric, pneumatic, or other types of power. Robotic grippers are usually tasked with interacting with their surroundings and gripping items by coming into direct contact with said workpiece [15, 16]. Making safe manipulation and detecting things being grasped are the primary needs for flexible grippers.

Industrial Internet of Things (IIoT) and its allied fields, such as big data, cloud computing, and Industrial Wireless Networks (IWNs), have made significant strides in recent years [8]. The arrival of the fourth industrial revolution, often known as Industry 4.0, will be made possible by these coming technologies, which will also present enormous prospects for boosting industrial improvements [5, 17, 18]. In order to comprehend how a hexapod animal walks numerically, the study proposes a model [19]. The goal is to create an appropriate hexapod walking robot biomimetic model that is smaller in weight, has fewer motors, and is simple to control. SOLIDWORKS is used to adapt design modifications. The moment and force analyses are then performed along with a thorough study of the model. Moment analysis is used to test the legs' ability to walk in a variety of extreme joint situations. The proposed model is determined to be suitable to endure varied forces while walking to execute any required task after the robot leg motion is tested by introducing weights at which the robot turns to fail [13, 19].

Some researchers presented a comprehensive assessment with practical applications for mobile robot stages that depend on certain wheels. In practically every area of mobile robot technology today, including flying robot, swarm robotics, and humanoid applications, omni-directional wheels are in demand. Due to their adaptability, small size, low price, and endurance, Omni-Directional Mobile Robots (OMR) with mecanum wheels are most frequently chosen among them [20]. For a robot to function better, it's crucial to understand its many parameters and qualities. To comprehend the dynamics of the robot needed for design and control, the inverse and forward kinematics are crucial [21].

Biped climbing robots are capable of manoeuvring and doing the job within 3-dimensional steel structures, but they are very dependent on the size, weight, and adsorption capabilities of the objects connected to both of their ends. Here, researchers suggest a unique gripper with the upper layer being able to spin coaxially with regard to the lower layer [22]. A straightforward and reliable robotic hand created for industrial purposes is the Adaptive Gripper. It features three fingers that can adjust to the form of different things and hold onto them. A gripper's primary function is to impart forces and torques to the item that has to be gripped. Grip force, adhesion between the gripper and the item, and geometric fit between the grippers and the contact surfaces of the object are all factors that affect how stable and reliable the gripping process is. For all items and circumstances, these factors cannot be maximized, especially if the design is constrained by weight and space constraints. These limitations result in straightforward, lightweight grippers, but they are typically inappropriate since they lack the ability to adapt to various forms. Complex grippers requiring a lot of input data and a high number of activated Degrees of Freedom (DOF) are difficult to regulate.

All odd-shaped components that make up the assembly's Body-in-White (BIW) are moved, positioned, and connected using a robot gripper in the prototype production cell before being welded or bolted together to create the assembly's final geometry. The whole handling procedure is included in the manufacturing cell. These grippers are often constructed around a gripper base, which is typically composed of plastic or steel. Base clamps, locators, and adjustment boxes are all added to this gripper since they are required to accommodate the ensuing assembly processes. Such grippers are rather heavy, which limits the amount of labor that can be carried by a robot of a certain size [23]. The present robotic equipment encountered a variety of issues throughout the transportation and alignment of the cast metal components, including relating to mass inertia, precision, stability, and the combined weight of the gripper and sheet metal portion. Either more potent robots needed to be added to the manufacturing system, or lighter fixtures needed to be created. The latter option was chosen because it was less expensive to acquire and because it was the only robot that could be fitted with a fully integrated tooling device.

Additionally, the new grippers should be flexible enough to fit existing and future metal components, enhance assembly operations' tolerances, remove or greatly minimize the requirement for regular calibrations, and boost the ratio of production capacity. The grippers needed to be lightweight, and a technology that would also provide the necessary stiffness to meet the tolerance criteria was selected: 3D printing composite [24]. Composite, a material that can be produced by 3D printing, has unique qualities that make fixture design quite difficult. Additionally, because it is so difficult to interface these pieces with other components, it is exceedingly challenging to construct modular (flexible) fixtures for 3D printing composite. Therefore, the only option was specialized fixtures.

The goal of this work is to develop and create a new module for Fable that will allow it to recognize common household goods. The gripper will be deemed adequate if it is able to pick up at least nine out of the ten objects shown in Fig. 1 without harming them. Different forms (spherical, cuboid, cylindrical, irregular…), materials (plastic, metallic, cloth, ceramic…), sizes, and weights will all need to be handled by the gripper [25]. Titanium is the fourth most abundant structural metal after aluminium, iron, magnesium and ninth most plentiful element [1, 2]. Titanium is available as commercially pure and as alloys. These alloys are usually divided into three categories alpha (α), alpha–beta (α + β) and beta (β). The α + β alloys, the most widely used alloy group, contain one or more α stabilizers plus one or more β stabilizers. Amongst the α + β alloys, Ti-6Al-4V and its variants are by far the most popular titanium alloys and constitute more than 50% of all titanium alloy applications [1].

In this work, MA Ti-6Al-4V ELI titanium alloy has been subjected to HT above beta transus using a vacuum hardening furnace. It was followed by aging treatment. The microstructure and mechanical properties of MA and β annealed titanium alloy have been investigated and compared.

Fig. 1 Nine items that fable should grasp: 0.5 L bottle, can, egg, shoe, orange, cardboard box, Fable's brick, cup, marker and teddy

All user requirements must be taken into consideration throughout the design phase of the Fable system in order to fulfil its intended function. In addition to the significance of aesthetics, it is also important to take into account the cost of the manufacturing process, its complexity, and the price of its components as it is anticipated to be marketed at some point.

2 Design and Development of Robotic Gripper

The design and development process for an adaptive gripper is described below.

2.1 MG995 Servo Motor

The MG995 servo is a straightforward, frequently employed basic servo for any mechanical requirements, such as a robotic head or arm. It has metal gears for maximum torque and a standard 3-pin control and power cable for simplicity of use. MG995 Servo is shown in Fig. 2.

Fig. 2 MG995 servo

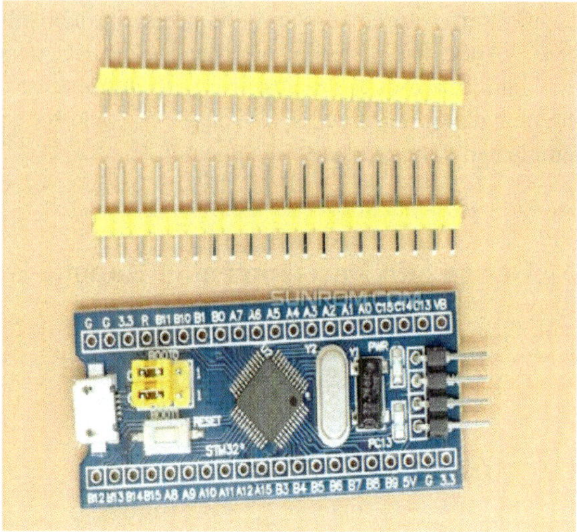

Fig. 3 STM32

2.2 *Microcontroller (STM32)*

The STM32 family of 32-bit microcontroller integrated circuits is manufactured by STMicroelectronics. Similar series among the STM32 processors that are constructed around the same 32-bit ARM processing core include the Cortex-M33F, Cortex-M7F, Cortex-M3, Cortex-M0+, and Cortex-M0. From basic printers to intricate circuit boards in automobiles, these microcontrollers are utilized in a variety of applications. Therefore, technical expertise in creating embedded systems and firmware using STM32 microcontrollers is a necessary skill set for an engineer in electronics and communications. STM32 is seen in Fig. 3.

2.3 *Power Regulator LM2596S*

The buck-switching regulators in the LM2596S series, which have good line and load regulation and can drive a 3-A load, are monolithic integrated circuits that perform all of the active functions. LM2596S is depicted in Fig. 4.

2.4 *Arduino Nano*

The Arduino Nano, which is based on the ATmega328, is a small, comprehensive, and breadboard-friendly board (Arduino Nano 3.x). It is packaged differently but has about the same capabilities as the Arduino Duemilanove. It only lacks a DC

Fig. 4 LM2596S

Fig. 5 Arduino Nano

power connector and operates using a Mini-B USB cable rather than a standard one. Figure 5 depicts an Arduino Nano.

2.5 USBasp

A USB in-circuit programmer for Atmel AVR controllers is called USBasp. Figures 6 and 7 depict the USBASP and device layout, respectively.

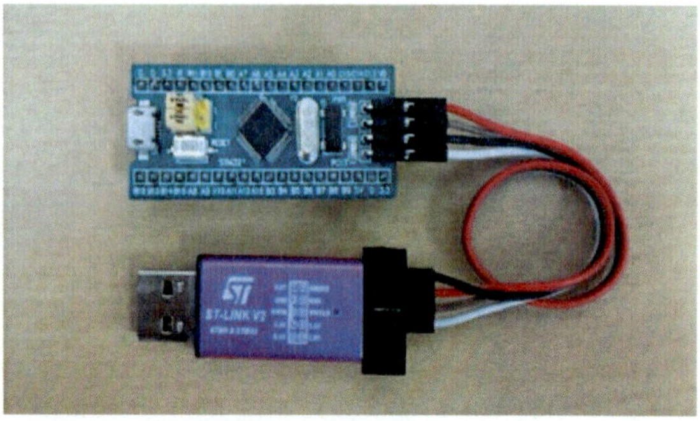

Fig. 6 Arduino Nano

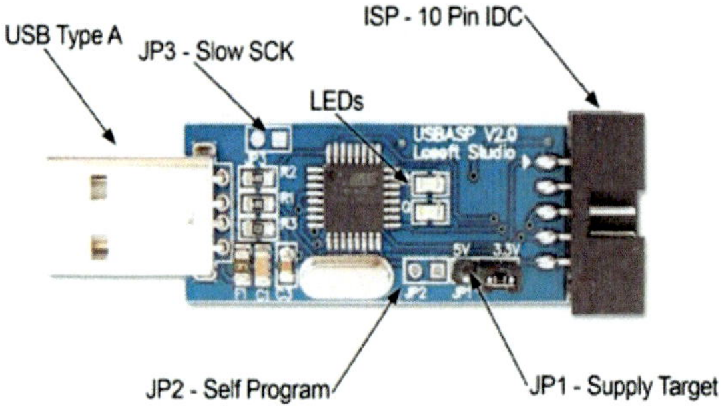

Fig. 7 Device layout

3 Manufacturing of Adaptive Gripper

A physical object is created from a computer design using the additive manufacturing process of three-dimensional (3D) printing. Thin layers of material, such as liquid or powdered plastic, metal, or cement, must initially be placed down for the process to work. The layers must then be fused together.

3.1 Ender 3 V2 3D Printer

A print bed with a tempered glass covering and a textured surface is a feature of the Creality Ender-3 V2. It can reach temperatures of up to 100 °C (212 °F), which expands the spectrum of printed materials, enhances layer adhesion, and makes print removal easier. Ender 3 V2 3D Printer is seen in Fig. 8.

The 3D-printed fingers are shown in Fig. 9 with the entire assembly. Due to several constraints, adjustments in the design were made afterwards. The redesigned design overcame the drawbacks.

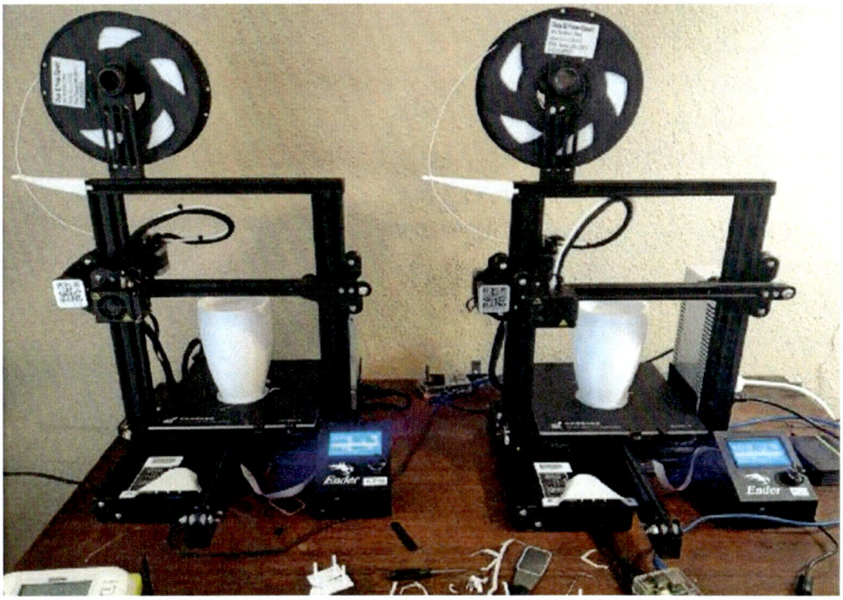

Fig. 8 Ender 3 V2 3D printer

Fig. 9. 3D printed gripper

The 3D-printed fingers are shown in Fig. 9 with the entire assembly. Due to several constraints, adjustments in the design were made afterwards. The redesigned design overcame the drawbacks.

The gripper's 3D-printed components, including the gripper base and adaptable grippers, are shown in Fig. 10. Figure 11 shows 3D-printed parts of the modified Gripper assembly.

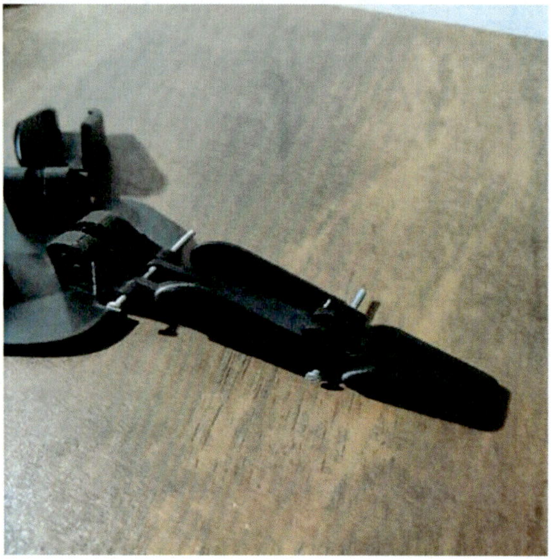

Fig. 10. 3D printed parts of modified gripper

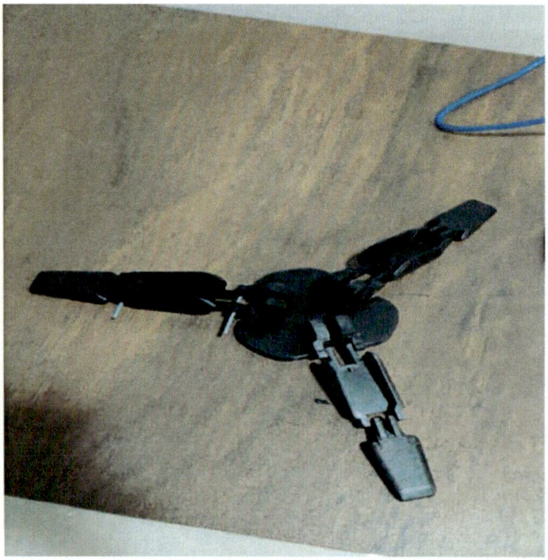

Fig. 11. 3D-printed parts of modified gripper assembly

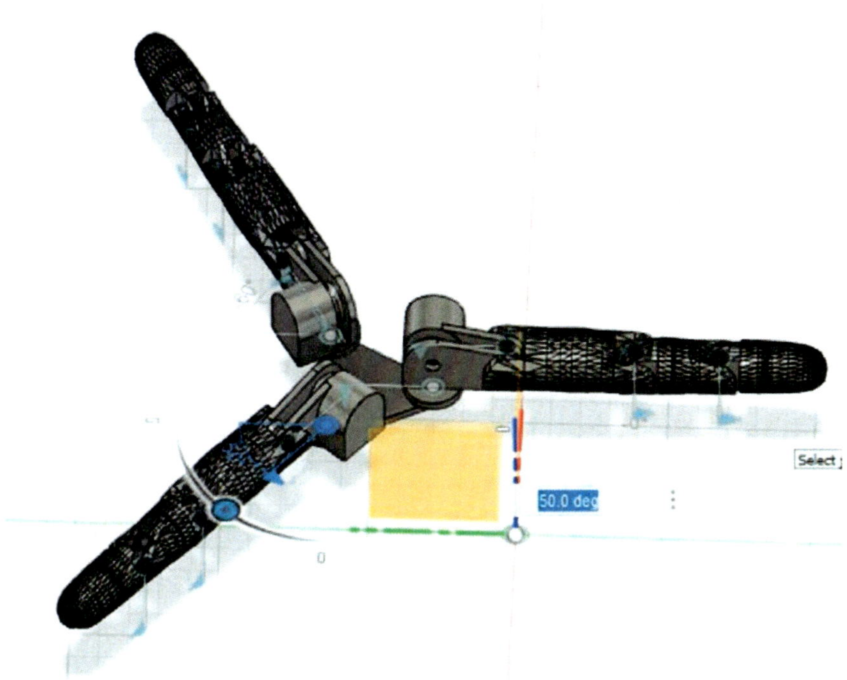

Fig. 12 Gripper design using fusion 360

3.2 Modelling of Adaptive Gripper

Fusion 360 was used to model the adaptive gripper, as shown in Fig. 12. Further, Figs. 13 and 14 show the gripper assembly and modified gripper design with modification respectively.

4 Fea Simulation of Adaptive Gripper

The von Mises stress and contact pressure in the gripper are calculated using the finite element programme "FUSION 360". As can be seen, the maximum von Mises stress for the material is 14.94 MPa, which is considerably less than the maximum Von Mises stress that is allowed. The von Mises stress distribution is seen in Fig. 15. The von Mises stress is a number used to predict whether a material will yield or fracture. Its primary uses are in metals and other ductile materials. A material will yield if its von Mises stress under load is the same as or larger than its yield limit under simple tension, according to the von Mises yield criterion.

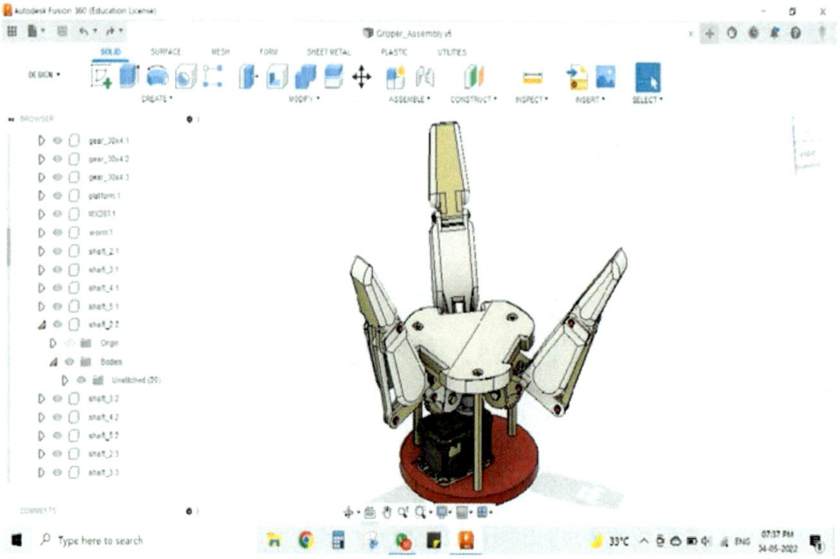

Fig. 13 Gripper assembly using fusion 360

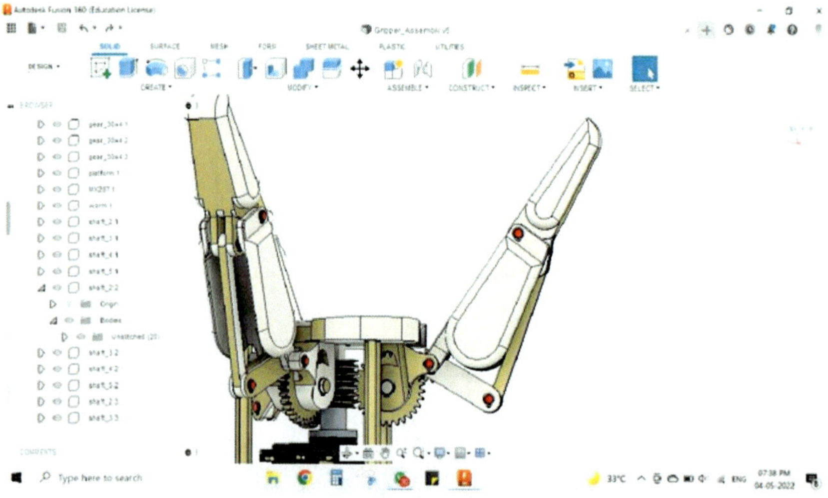

Fig. 14 Modified gripper design with mechanism

For a finite element model created in this study, total deformation and directional deformations are shown in Fig. 16. When you consider the boundary conditions that were imposed on the gripper, the overall deformation of 0.2138 mm is quite little. The displacement of the system in a certain axis or user-defined direction is referred to as

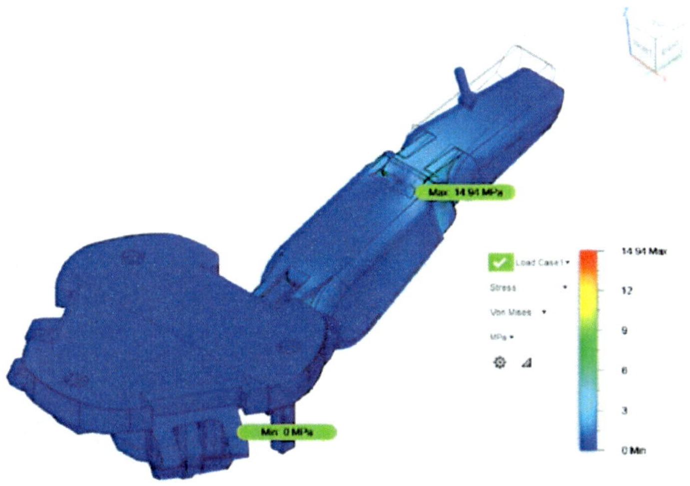

Fig. 15 von Mises stresses

directional deformation. The vector sum of each system's directional displacements represents the total deformation.

In response to applied loads, the gripper created directional responses and consequent reaction forces, which are shown in Fig. 17.

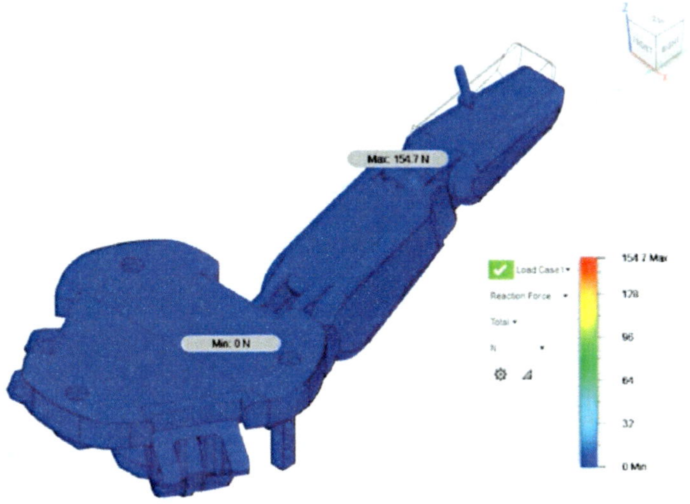

Fig. 16 Total deformation

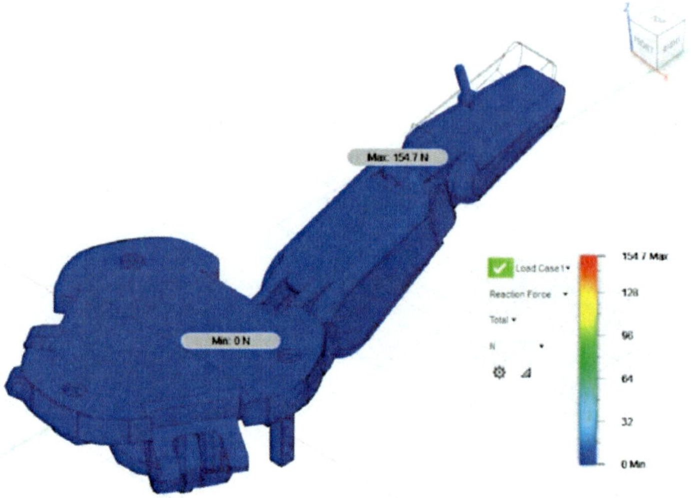

Fig. 17 Reaction forces

Under the given loading and boundary circumstances, Fig. 18 shows the lowest and maximum values of the gripper's induced strains as well as any additional strains (see Fig. 19).

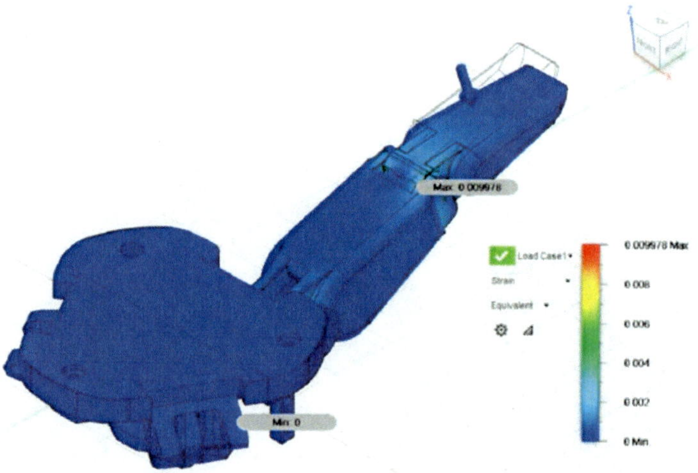

Fig. 18 Equivalent strain

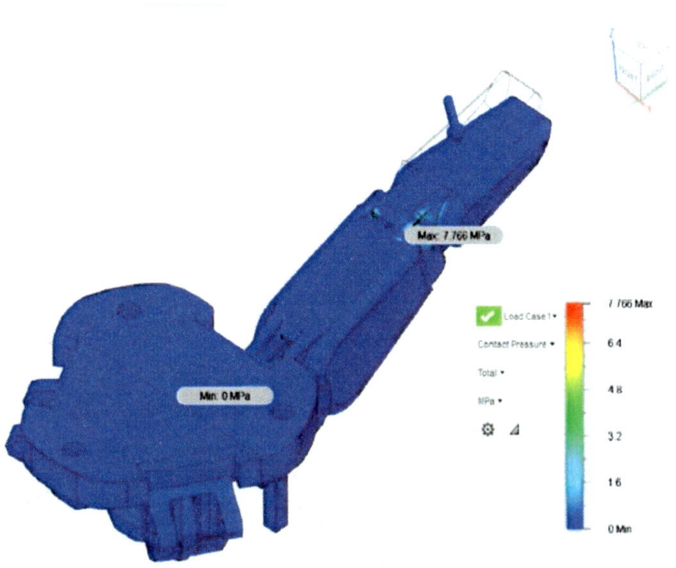

Fig. 19 Contact pressure

5 Conclusions

This study was the first attempt to design and create an adaptable gripper, although it was not without constraints. Following are the concluding remarks from the present study.

1. The cable stretched after a specific number of duty cycles, which was the limiting. Furthermore, the finger design proved insufficient for grabbing spherical items. In the current work, this constraint is solved by altering the existing gripper design.
2. Using 3D printing technology, a low-cost adaptable gripper was created. In the creation of this adaptable gripper, two designs were investigated.
3. In the original design, some technical issue was faced with the wire extending while working. As a result, it was intended to make design improvements such as switching cables to links.
4. The redesigned design eliminates rotary servos 1 and 2, lowering the total cost of the adaptive gripper compared to its predecessor.

References

1. Patil LN, Khairnar HP (2021) Investigation of human safety based on pedestrian perceptions associated to silent nature of electric vehicle. Evergreen 8:280–289. https://doi.org/10.5109/4480704
2. Patil LN, Khairnar HP (2021) Assessing impact of smart brake blending to improve active safety control by using Simulink. Appl Eng Lett 6:29–38. https://doi.org/10.18485/aeletters.2021.6.1.4
3. Patil LN (2021) Assessment of risk associated with the quiet nature of electric vehicle: a perception of EV drivers and pedestrians at Mumbai metropolitan region-India. Eur Transp/Trasporti Europei 1–19. https://doi.org/10.48295/ET.2021.83.1
4. Galabov V, Stoyanova Y, Slavov G (2014) Synthesis of an adaptive gripper. Appl Math Model 38:3175–3181.https://doi.org/10.1016/j.apm.2013.11.038
5. Patil AA, Desai SS, Patil LN, Patil SA (2022) Adopting artificial neural network for wear investigation of ball bearing materials under pure sliding condition. Appl Eng Lett 7:81–88. https://doi.org/10.18485/aeletters.2022.7.2.5
6. Patil LN, Khairnar HP, Hole JA, Mate DM, Dube AV, Panchal RN, Hiwase VB (2022) An experimental investigation of wear particles emission and noise level from smart braking system 09:10
7. Patil L, Khairnar H (2021) Enhancing safety of electric vehicle drivers through vision oriented monitoring system: vision oriented monitoring system. In: Advances in robotics—5th international conference of the robotics society. ACM, Kanpur India, pp 1–5
8. Patil L, Khairnar H (2022) Python inspired smart braking system to improve active safety for electric vehicles. IJAME 19:9447–9459. https://doi.org/10.15282/ijame.19.1.2022.08.0727
9. Ruiz-Ruiz FJ, Ventura J, Urdiales C, Gómez-de-Gabriel JM (2022) Compliant gripper with force estimation for physical human–robot interaction. Mech Mach Theory 178:105062. https://doi.org/10.1016/j.mechmachtheory.2022.105062
10. Lien TK (2013) Gripper technologies for food industry robots. In: Robotics and automation in the food industry. Elsevier, pp 143–170
11. Hou Z, Li Z, Fadiji T, Fu J (2021) Soft grasping mechanism of human fingers for tomato-picking bionic robots. Comput Electron Agric 182:106010. https://doi.org/10.1016/j.compag.2021.106010
12. Kang L, Seo J-T, Kim S-H, Kim W-J, Yi B-J (2019) Design and implementation of a multi-function gripper for grasping general objects. Appl Sci 9:5266. https://doi.org/10.3390/app9245266
13. Xiao W, Liu C, Hu D, Yang G, Han X (2022) Soft robotic surface enhances the grasping adaptability and reliability of pneumatic grippers. Int J Mech Sci 219:107094. https://doi.org/10.1016/j.ijmecsci.2022.107094
14. Petković D, Seyed Danesh A, Dadkhah M, Misaghian N, Shamshirband S, Zalnezhad E, Pavlović ND (2016) Adaptive control algorithm of flexible robotic gripper by extreme learning machine. Robot Comput-Integr Manuf 37:170–178https://doi.org/10.1016/j.rcim.2015.09.006
15. Zhang B, Xie Y, Zhou J, Wang K, Zhang Z (2020) State-of-the-art robotic grippers, grasping and control strategies, as well as their applications in agricultural robots: a review. Comput Electron Agric 177:105694. https://doi.org/10.1016/j.compag.2020.105694
16. Kaviyarasan S, Infanta Mary Priya I (2018) Design and fabrication of three finger adaptive gripper. In: IOP conference series: materials science and engineering, vol 402, p 012043. https://doi.org/10.1088/1757-899X/402/1/012043
17. Liu C, Cheng J, Li Z, Cheng C, Zhang C, Zhang Y, Zhong RY (2020) Design of a self-adaptive gripper with rigid fingers for Industrial Internet. Robot Comput-Integr Manuf 65:101976. https://doi.org/10.1016/j.rcim.2020.101976
18. Mishra A, Priyadarshini R, Choudhary S, Mehra RM (2021) Qualitative analysis of intra-class and inter-class clustering routing and clusterization in wireless sensor network. Evergreen 8:358–373. https://doi.org/10.5109/4480718

19. Shrivastava V, Diwakar V, Sehgal M, Verma M, Neha E (2022) Modelling and analysis of hexapod walking robot. Evergreen 9:378–388. https://doi.org/10.5109/4794162
20. Yadav PS, Agrawal V, Mohanta JC, Ahmed MDF (2022) A theoretical review of mobile robot locomotion based on mecanum wheels. Evergreen 9:396–403. https://doi.org/10.5109/4794163
21. Chauhan SS, Khare AK (2020) Kinematic analysis of the ABB IRB 1520 industrial robot using Robo analyzer software. Evergreen 7:510–518. https://doi.org/10.5109/4150470
22. Zhu H, Lin Z, Yan J, Ye P, Zhang W, Mao S, Guan Y (2022) Compact lightweight magnetic gripper designed for biped climbing robots based on coaxial rotation of multiple magnets. Robot Auton Syst 155:104164. https://doi.org/10.1016/j.robot.2022.104164
23. Andronas D, Xythalis S, Karagiannis P, Michalos G, Makris S (2021) Robot gripper with high speed, in-hand object manipulation capabilities. Procedia CIRP 97:482–486. https://doi.org/10.1016/j.procir.2020.08.007
24. Cheng L-W, Chang J-Y (2018) Design of a multiple degrees of freedom robotic gripper for adaptive compliant actuation. In: 2018 international conference on system science and engineering (ICSSE). IEEE, New Taipei, pp 1–6
25. Telegenov K, Tlegenov Y, Hussain S, Shintemirov A (2015) Department of robotics and mechatronics, school of science and technology, Nazarbayev university, school of mechanical, materials and mechatronic engineering, university of Wollongong: preliminary design of a three-finger underactuated adaptive end effector with a breakaway clutch mechanism. J Robot Mechatron 27:496–503. https://doi.org/10.20965/jrm.2015.p0496

An Innovative Review on the Oscillating Drum Ball Milling Process

Devesh Mishra, Abhishek Manohar, Mohit Pal, Kritagya Chaudhary, Nikita Nath, Nitin Kumar, Girendra Bhati, Iqbal Ahmed Khan, and Syed Qaisar Husain

Abstract In the present scenarios, ceramic matrix composites (CMCs) have been mixed under the action of ball milling operations. Generally, the ball milling processes have been implemented by either rotation or reciprocating of the mixing drum. In a rotating type mixing drum, the spherical steel or ceramic balls rotate inside the cylinder under the action of centrifugal forces and generate the problem of sticking the materials on the inner circular wall sections. The problem with the reciprocating mixing drum is that the spherical balls are sedimented in the solution under the action of gravity force. In both cases, the non-uniform and heterogeneous phases of slurry have been observed. The main objective of this research work is to eliminate these critical issues through the concept of kinematics and mechanism, named as oscillating drum ball milling process.

Keywords Nano-ceramics · Reinforcement · Oscillating drum · CMCs

Nomenclature

CMCs	Ceramic Matrix Composites
Cr-YAG	Chromium-Yttrium Aluminum Garnet
YSZ	Yttria-stabilized Zirconia
ZTA	Zirconia Toughened Alumina
ZrO_2	Zirconia
SEM	Scanning Electron Microscope
EDS	Energy-dispersive X-ray spectroscopy
XRD	X-ray Powder Diffraction
TEM	Transmission Electron Microscopy

D. Mishra · A. Manohar · M. Pal · K. Chaudhary · N. Nath · N. Kumar · G. Bhati (✉) · I. A. Khan · S. Q. Husain
Department of Mechanical Engineering, Greater Noida Institute of Technology, Greater Noida 201310, India
e-mail: girendrabhatime@gmail.com

A. Sachdeva et al. (eds.), *Recent Advances in Operations Management and Optimization*, Lecture Notes in Mechanical Engineering,
https://doi.org/10.1007/978-981-99-7445-0_2

CNT Carbon Nano-Tubes

1 Introduction

Nano-ceramic patches are nano-patches made primarily of pottery. Nano-ceramic patches are heat-resistant, non-metallic, and inorganic solids composed of both metallic and non-metallic chemicals from which a wide range of mixes may be created. Some of the functionalities are handed by nano-composites. Composites include dielectric, ferroelectric, piezoelectric, pyroelectric, ferromagnetic, magnetoresistive, superconductive, and electro-optical. The first nano-ceramic flyspeck was discovered in the early 1980s. They were created exercising a sol–gel fashion, which combines nanoparticles in a result and gel to produce the nanoparticle. Sintering was used after procedures. Because the material is so bitsy, there are nearly no faults. Larger scale accouterments are fragile due to blights. This construction may recapture its original shape after repeated smashing. Ceramic nanoparticles have been employed as a medicine delivery system in a variety of affections, including bacterial infections, glaucoma, and chemotherapy administration in experimental cancer treatment. Nano-tubes are ground into very tiny powders using the ball milling method. This method is mostly employed to mix ingredients and finely ground powders. This procedure is cost-effective, kind to the environment, and has found widespread use in business throughout the globe. Typically, it comprises a hollow cylindrical tank revolving around its axis that contains rubber, steel, stainless steel, ceramic, or glass balls. Reliability, simplicity of use, consistent outcomes owing to energy and speed control, and application in wet and dry circumstances on a variety of materials, including hydroxyapatite, are benefits of this sort of technology.

In the ball milling process, the nanotubes are ground into incredibly tiny particles. Ball milling can damage the nanotubes even if it is simple to use and appropriate for powder polymers or monomers. Ball milling is frequently used for the creation of oxides, nanocomposite materials, and/or the improvement of structure/phase composition in addition to grinding powders. It is a mechanical process often used to create small particles from powders. By randomly impacting graphite (the precursor) with grinding media inside a revolving shell, this technique broke down the precursor, creating shear and compression force that helped overcome the weak Vander Waal's contact between the graphite layers and caused their splintering. Cost-effectiveness, dependability, simplicity of use, consistent results due to energy and speed control, and applicability in wet and dry circumstances on a variety of materials are all benefits of this technology. Moreover, there are certain drawbacks such as the potential for contamination, the development of nanomaterials with irregular shapes, noise, and lengthy grinding.

2 Literature Review

Guo et al. studied the dependency of rheological characteristics of slurries, along with the reaction of hydration of oxide powders all along composing translucent Cr: YAG ceramics depreciated $Cr_2O_3/Al_2O_3/Y_2O_3$ powders with the help of slip casting and sintering of vacuum. They identified that the dispersant can feed the Y and Al group which has been blended in the slurry mixture and immersed on the outside of the uninvolved powders of oxides and it shifts the pH and rheological characteristics of slurry, consequently, another involves the means of slip casting and the sintering of the ceramic. They gained that Cr: YAG ceramic by the common grain size of ~6.2 μm and in-line transmittance is more than 84% at 1450 nm later sintering the green samples in 1800 °C under vacuum conditions for 40 h [1]. **Sabzi et al.** investigated metal matrix composites for extreme environmental conditions and service pressures. They conclude that the electroplating process is one of the acceptable ways that have been examined by numerous researchers in the past due to its simplicity, low cost, and economical manner of developing nanocomposite coatings. They investigated numerous aspects such as the concentration of various additives and lubricants, temperature and current density, frequency of current and stirring speed, roughness, and other mechanisms. They determined that the overall process of ceramic or metal deposition consisted of two stages: nucleation and growth. They discovered that changing the pH of the electroplating bath causes metal ion deposition, contaminants to rise, and residual stresses to increase in nanocomposite coatings [2]. **Li et al.** studied the enhancement of the automated settings of excellent ceramics manufactured by layer-based manufacturing is essentially related to the broad function. In this study, a range of yttria-stabilized zirconia (YSZ) materials with zirconia toughened alumina (ZTA) capacity were satisfied and manufactured through optical fabrication techniques planted on the self-holding capacity of the viscoelasticity fix and function of adaptable contact base method. Linking microstructural investigation and automatic appraisal by three-point curving and impression analysis, the effect of YSZ on the SL procedure and consecutive automatic characteristics was explained. Hence, the outcome shows that the advanced corresponding coil power (477.54 MPa) and fracture grit (6.56 MPa $m^{1/2}$) were taken by ZTA ceramic with a YSZ content of 20 wt.% [3].

Wachowski et al. studied Ti/Al_2O_3 ceramic metal compound gained with the help of slip-casting process fragments overall 50 volumes of the solid state along with 10 volumes of the metal state were checked the rheological prospects which concluded a thermogravimetric study achieved the characteristics of gained compound resolved the step formation by applying SEM/EDS micro-structural study the approach of XRD of the size of titanium particles equivalent to 20.6 ± 101 mm which coincides with 275 of original capacity [4]. **Fatemeh et al.** observed that the balanced slurries were the distributed variety of popular Al_2O_3 and Y_2O_3 granules, ammonium poly meta-acrylate (Dolapix CE-64) as the Dissipate, and tetraethyl orthosilicate (TEOS) as the heating aid were made by ball milling method. The outcome of the time of milling on the fabrication of clear polycrystalline yttrium aluminum garnet (YAG) ceramics was inspected by slip casting and vacuum sintering. The outcomes show that the perfect milling time for the deagglomeration of the powder blend was 16 h and the slurry adapted at this time shows a near-Newtonian action due to the superior deagglomeration and low viscosity. radiography arrangements also produced all patterns in pure YAG state. The outcome also exposed that the patterns prepared by 16 h of ball milling time break presented more relative green and finishing density and maximal change at 1064 nm ($\approx$77%) [5].

Bechteler et al. delved carbon nanotube (CNT) corroborated alumina nanocomposite They discovered that grain size has a considerable impact on the percolation threshold of electrical conductivity, performing in anisotropic characteristics. They also discovered that when CNT attention increases, thermal conductivity falls vastly. They determined that the electrical conductivity of CNT/alumina mixes is dependent on the main greasepaint flyspeck size and that an increase in CNT content increases porosity, which imposes anisotropic electrical conductivity. They concluded to the conclusion that characterizing the independent impact of Carbon nanotube content and porosity on thermal conductivity revealed that adding Carbon nanotube content at constant porosity, as well as adding porosity at constant Carbon nanotube, results in a considerable drop in thermal conductivity [6]. **Idrees et al.** fitted a ball milling process, delved Silicon boron carbon nitride (SiBCN) ceramic mixes filled with nitrogen sulfur binary-doped graphene (NSG) distance (SiBCN/NSGs) were cooked and manufactured. They discovered that NSGs are foreign excrescencies that increase electrode performance and have a reversible capacity of 785 mAh. They conclude that the goods of nitrogen sulfur binary-doped graphene (NSG) distance active anode material was attained through pyrolysis at 600, 800, and 1000 °C and that these unique silicon-grounded carbon accouterments have great eventuality to be used for other operations similar to electromagnetic hindrance [7]. **Song et al.** delved UV-curable and high-density ZrO_2 ceramic nanocomposite resins. They employed APTMS-carpeted ZrO_2 ceramic patches with di and tri-functional acrylate monomers with non-reactive diluents, grounded on a percolating network with 50 ceramic patches and a high density, to increase their dissipation stability and print-curing cap behavior abilities. At 1450 °C, the green bodies of 3D-published objects with varying cross-linking consistency were sintered. Grounded on the results of density and UV curing rheological and print-curing behaviors, they conclude that IPA attention was stylish at 20 monomers [8].

Asl et al. explored the mongrel ZrB_2 a grounded compound with 10 nano-sized carbon dark and 20 SiC and manufactured by vacuum hot pressing for 60 twinkles at 1850 °C under 20 MPa. The sintered ceramic's structure and sintering ability were delved into using X-ray grease paint diffraction, surveying electron microscopy, and X-ray spectroscopy. They found that no conspicuous in situ-generated new phases were observed at the graphite ZrB_2 and SiC grain contact. Even though the densification process was chemically inert, some of the unformed carbon black patches changed into liquid graphite nano-flakes during hot pressing [9]. **Eatemadi and Balakl** analyzed the influence of sintering settings on the fracture durability and densification of spark tube sintering nanocomposites. They sintered ZrB_2 30% vol

SiC nanocomposites at 1600–1800 °C for 8–12 min. They delved into the microstructural differences using scanning electron microscopy (SEM). They discovered that raising the sintering temperature redounded in a minor increase in SiC grain size but no sensible variation in ZrB_2, and that adding the sintering duration redounded in an increase in both ZrB_2 and SiC grain size. They concluded that analysis of SEM images and path propagation revealed that crack divagation and branching are asked toughening processes in ZrB_2–SiC nanocomposites [10]. **Lu et al.** delved free-standing and flexible polymer-ceramic nanocomposite flicks and created them using a spin-coating procedure using poly (vinylidene fluoride-co-hexafluoropropylene) as the matrix and $BaTiO_3$ (BTO) nanoparticles as the padding. They discovered that adding the BTO attention reduces crystallinity and demitasse size in the P (VDF-HFP) matrix. They found that dielectric responses produced in compound flicks are affected by temperature and frequency, and the goods of the ceramic padding on the four dielectric processes of the polymer matrix were delved. They find that when the BTO attention increases, so do the crystallinity and size of the P(VDFHFP) matrix chargers. Because of the high r and low tan of BTO padding, the r increases and the tan drops with adding BTO attention at room temperature [11].

Wagih et al. carpeted Ag nanoparticles with mongrel Cu-Al_2O_3 nanocomposites were delved, and new mixes with bettered mechanical, electrical, and thermal parcels were developed. Compressive strength, hardness, thermal, and electrical parcels were also delved. They discovered that adding GNPs improves compressive strength and hardness by 88.1 and 55.2, independently [12]. **Ghayour et al.** used different types of scenarios that increase the BPR such as weight loss of powder, change of diameter, and the number of balls and their effects on the milling efficiency. They noticed that results show the loss of powder weight can greatly enhance the performance of milling. When the quantity of balls is increased at a high BPR ratio, it influences the milling performance which is negative. They obtained results, if the vial-to-plate spinning rate is 1.2, however by an increase in diameter the mill had a good performance, and this ratio is changed to 1.4 [13]. **Jayaramudu et al.** used nanoparticles of silver with a size of crystallite range between 10 and 12 nm, synthesized poly (ethylene glycol) which is of different molecular weights as agents [14]. **Song et al.** prepared the composite of CNTs/Cu through the electroless plating and planetary ball milling process. They noticed that the results show that copper particles are deposited homogeneously on carbon nano-tubes surface. It was discovered that when comparing composite with electroless plating to composite without, the tensile strength of the latter improves noticeably, up to 207 MPa. The cold drawing was successful in producing composite materials with greater tensile strength. The mechanical characteristics of the composites have been increased, according to the results. The tensile strength and yield strength are respectively up to 311 MPa and 185 MPa [15]. **Zhang et al.** show graphene as the ideal reinforcement because of its favorable mechanical and physical characteristics. They employed milling balls with a tiny diameter and strengthened graphene nanoplates to minimize the impact energy during the ball milling process. Ball milling was used to successfully create composite materials made of the Al-5083 alloy. By using X-ray diffraction, the structures of the micro-level ball-milled powders and GNP/Al-5083 composites were identified.

The outcomes demonstrated that the GNP persisted after ball milling. While some GNP may have interacted with Al to generate Al_4C_3 during the merging process, the Al_4C_3 phase was discovered in bulk GNP/Al-5083 composites [16]. **Oqail et al.** used copper nanocomposites assisted by 5 and 10 wt.% of ZrO_2 compound particles. Nanocomposite powder which is produced was examined by X-ray diffraction and transmission electron microscopy (TEM). The produced powders were then subjected to cold compaction at a pressure of 700 MPa and sintered for two hours at 950 °C in a hydrogen environment. This demonstrates that increasing the milling time increases the produced nanocomposites' microhardness. After 20 h of milling, Cu-10%ZrO_2 has 3.76 times the microhardness of pure Cu [17]. **Wirunchit et al.** noticed that the size and morphology of ZnO nanoparticles varied depending on the calcination temperature, which ranged from 400 to 900 C. It was investigated that high-purity ZnO is obtained around 800 °C [18].

Ramezanalizadeh et al. investigated how ball milling altered the nanocomposite powder made of aluminium's microstructure. They noticed that when milling time and lattice strain increased, the crystal size of nanocomposite powders decreased. He concluded that 15 h of milling produced a homogeneous distribution of Al_3Mg_2 in the matrix of Al. He also saw a significant size reduction in the Al_3Mg_2 particles [19]. **Caicedo et al.** prepared graphene oxides (GO) by ball milling them with potassium perchlorate, deionized water here is known as milling media. The sample that was processed for 12 h had the best yield of the product with the desired particle size (40 nm) for a variety of biological applications, they found. As a result, they discovered that the ball milling method for making graphene is ideal for use in medical applications, while SAED pattern analysis revealed the sample was crystalline [20]. **Kumar et al.** found that the mechanochemical production of nano-biochar is dominated by the ball milling method. They saw that this method was inexpensive and environmentally beneficial. They hypothesized that the creation of biochar nanoparticles with certain desirable characteristics, such as 400–500 m^2 g1 surface area and 0.5–1000 nm pore diameters, opens up a variety of applications in energy. They found that some chemical properties and uses of ball-milled biochar are largely influenced by the biochar feedstock and ball-milling settings. The process used to create the higher-quality biochar-based material is completely environmentally friendly and economical [21]. **Singh et al.** used the ball milling method and four different milling mediums to create Zno (DI water, ethanol, isopropanol). They also looked at how different media affected the structural and optical characteristics of the ZnO nanoparticles. They examined the ZnO nanoparticles using the X-ray diffraction method. They discovered that various milling media had an impact on structural properties. They discovered that the values of the band gaps of particles generated in various milling mediums have been determined to be in the range of 3.32–3.37 eV. Using various milling mediums, they saw that the crystallite size varied from 50 to 58 nm [22].

Maleki and Jashni examined how the local natural clay's characteristics will be affected by the ball milling procedure. The weight ratio of the balls to the powder was 10:1 or 20:1 as local clay was processed over a period ranging from 5 to 20 h. To determine the clay's particle size, researchers performed X-ray diffraction, Fourier transforms infrared spectroscopy, scanning electron microscope examination, and adsorption studies. They discovered that the amount of amorphization rose with grinding time, according to XRD measurements. They looked at adsorption studies, and this showed that, under comparable circumstances, BM-Clay had the highest capacity for adsorbing Ni (II) ions. At a pH value that is 7 and a temperature of 25 °C, they discovered that the Langmuir maximum adsorption capacity was 29.76 mg/g [23]. **Bor et al.** employed composite materials based on carbon nanotubes (CNTs). They employed the Planetary Ball Milling (PBM) method under several advantageous circumstances. To show the findings, they employed techniques including scanning electron microscope (SEM), and X-ray diffraction (XRD), in addition to field emission scan electron microscopy (FESEM). Using a high PBM rotation speed revealed that the CNTs were firmly lodged into the copper surface. On the other hand, copper particles with low PBM speeds showed practically little change. This indicates that the milling process's higher speed of rotation yields the best outcomes [24]. **Dash et al.** made use of graphene oxides (GO) produced by planetary ball milling with various forms of oxidation. They discovered that as the high-purity graphite sample was ball milled for longer periods of time, from two to twenty-four hours, the XPS examination revealed an increasing quantity of the atom concentration ratio of O/C. The GO sample created after 16 h of milling is ultimately deemed to be the most optimized sample in the presence of parameters including the degree of oxidation, duration, and energy consumption [25].

3 Problem Formulation and Future Scope

On the basis of the literature review, various types of ceramics and ceramic matrix composites are highly applicable in various industries such as orthopaedics, aerospace, automobiles, petroleum hydrotreatment, electronic insulators, fuel cells, and structures. There is a wide range of powder ceramic materials has been investigated such as alumina, nickel oxide, magnesium oxide, yttria-stabilized zirconia, silicon nitride, silicon carbide, bio-glass, hydroxyapatite, calcium phosphate, tricalcium phosphates, titanium oxides, yttrium silicate, phosphorus pentoxide, calcium oxide, barium titanate, ferric oxide, polyciliate, lead zirconate titanate, polycarbazole, divinyl benzene, LSCF ($La_{0.6}.Sr_{0.4}.Co_{0.2}.Fe_{0.8}.O_3$), CGO ($Ce_{0.9}.Gd_{0.1}.O_{1.95}$), etc. Many researchers introduced various binders and additives such as water, silica sol, camphene, naphthalene, oligomeric polyester, polyethylene glycol, ammonium polymethacrylate, glycerol, 4-hydroxy butyl acrylate, polyvinyl butyral, tert-butyl acrylate, tetraethyl-orthosilicate, nitric acid, isopropyl alcohol, gelatin, etc. CMCs have been mixed under the action of ball milling operations. Generally, the ball milling processes have been implemented by either rotation or reciprocating of the mixing drum. In a rotating type mixing drum, the spherical steel or ceramic balls rotate inside the cylinder under the action of centrifugal forces and generate the problem of sticking the materials on the inner circular wall sections. The problem with the reciprocating mixing drum is that the spherical balls are sedimented in the solution under the action of gravity force. In both cases, the non-uniform and heterogeneous phases of slurry have been observed. The main objective of this research is to eliminate these critical issues through a mechanism, named as oscillating drum ball milling process.

4 Conclusions

This review work represented the industrial applications of different types of ceramics as well as ceramics matrix composites, their mixing process, and manufacturing processes. It demonstrated the present and future aspects of ball milling operations. The improvement in the various mechanical as well as rheological properties such as density, viscosity, homogeneity, flowability, etc., may be a remarkable achievement in the mixing process of solution.

References

1. Guo W, Huang J, Lin Y, Huang Q, Fei B, Chen J, Wang W, Wang F, Ma C, Yuan X, Cao Y (2015) A low viscosity slurry system for fabricating chromium doped yttrium aluminum garnet (Cr: YAG) transparent ceramics. J Eur Ceram Soc 35:3873–3878
2. Sabzi M, Dezfuli SM, Balak Z (2019) Crystalline texture evolution, control of the tribocorrosion behavior, and significant enhancement of the abrasion properties of a Ni-P nanocomposite coating enhanced by zirconia nanoparticles. Int J Miner Metall Mater 26:1020–1030
3. Li M, Liu W, Nie J, Wang C, Li W, Xing Z (2021) Influence of yttria-stabilized zirconia content on rheological behavior and mechanical properties of zirconia-toughened alumina fabricated by paste based stereolithography. Ceramics 56:2887–2899
4. Wachowski M, Zygmuntowicz J, Kosturek R, Piotrkiewicz P, Żurowski R, Więcław-Midor A, Śnieżek L (2022) Study on manufacturing via slip casting and properties of alumina-titanium composite enhanced by Thialite phase. Material 16:79
5. Mohammadi F, Mirzaee O, Tajally M (2020) The effect of ball milling time on the rheological optical, and microstructural properties of YAG transparent ceramics. Appl Ceram Technol 17:1119–1127
6. Bechteler C, Machuj L, Hebendanz K, Rubling A, Girmscheid R, Kuhl H (2022) Electrical and thermal conductivity of CNT/alumina-nanocomposite ceramics. Ceram Eng Sci 5:2578–3270
7. Idrees M, Batool S, Kong J, Zhuang Q, Liu H, Shao Q, Lu N, Feng Y, Wujcik EK, Gao Q, Ding T, Wei R, Guo Z (2019) Polyborosilazane derived ceramics—nitrogen sulfur dual doped graphene nanocomposite anode for enhanced lithium ion batteries. Electrochim Acta 296:925–937
8. Song SY, Park MS, Lee D, Lee JW, Yun JS (2019) Optimization and characterization of high-viscosity ZrO_2 ceramic nanocomposite resins for supportless stereolithography. Mater Des 180:107960
9. Asl MS, Nayebi B, Kakroudi MG, Shokouhimehr M (2019) Investigation of hot pressed ZrB_2–SiC–carbon black nanocomposite by scanning and transmission electron microscopy. Ceram Int 45:16759–16764
10. Eatemadi R, Balak Z (2019) Investigating the effect of SPS parameters on densification and fracture toughness of ZrB_2-SiC nanocomposite. Ceram Int 45:4763–4770
11. Lu X, Zhang L, Tong Y, Cheng ZY (2019) BST-P(VDF-CTFE) nanocomposite films with high dielectric constant, low dielectric loss, and high energy-storage density. Compos Part B: Eng 168:34–43

12. Wagih A, Oqail AA, Fathy A (2019) Effect of GNPs content on thermal and mechanical properties of a novel hybrid Cu-Al_2O_3/GNPs coated Ag nanocomposite. Ceram Int 45:1115–1124
13. Ghayour H, Abdellahi M, Bahmanpour M (2016) Optimization of the high energy ball-milling: modeling and parametric study. Powder Technol 291:7–13
14. Jayaramudu T, Raghavendra GM, Varaprasad K, Reddy GVS, Reddy AB, Sudhakar K, Sadiku ER (2015) Preparation and characterization of poly(ethylene glycol) stabilized nano silver particles by a mechanochemical assisted ball mill process. J Appl Polym Sci 133:43027
15. Song JL, Chen WG, Dong LL, Wang JJ, Deng N (2015) An electroless plating and planetary ball milling process for mechanical properties enhancement of bulk CNTs/Cu composites. J Alloy Compd 720:54–62
16. Zhang H, Xu C, Xiao W, Ameyama K, Ma C (2016) Enhanced mechanical properties of Al5083 alloy with graphene nanoplates prepared by ball milling and hot extrusion. Mater Sci Eng: A 658:8–15
17. Oqail AA, Wagih A, Fathy A, Elkady O, Kabeel AM (2018) Effect of high energy ball milling on strengthening of Cu-ZrO_2 nanocomposites. Ceram Int 45:5866–5875
18. Wirunchit S, Gansa P, Koetniyom W (2021) Synthesis of ZnO nanoparticles by Ball-milling process for biological applications. Mater Today: Proc 47:3554–3559
19. Ramezanalizadeh H (2021) Effect of milling process on microstructural and properties of a ball-milled Al-based nanocomposite powder. Adv Mater Process Technol 8:2184–2192
20. Caicedo MCF, López EV, Agarwal A, Drozd V, Durygin A, Hernandez AF, Wang C (2020) Synthesis of graphene oxide from graphite by ball milling. Diam Relat Mater 109:108064
21. Kumar M, Xiong X, Wan Z, Sun Y, Tsang DCW, Gupta J, Gao B, Cao X, Tang J, Ok YS (2020) Ball milling as a mechanochemical technology for fabrication of novel biochar nanomaterials. Bioresour Technol 312:123613
22. Singh J, Sharma S, Soni S, Sharma S, Singh RC (2019) Influence of different milling media on structural, morphological, and optical properties of the ZnO nanoparticles synthesized by ball milling process. Mater Sci Semicond Process 98:29–38
23. Maleki S, Karimi-Jashni A (2017) Effect of ball milling process on the structure of local clay and its adsorption performance for Ni(II) removal. Appl Clay Sci 137:213–224
24. Bor A, Ichinkhorloo B, Uyanga B, Lee J, Choi H (2018) Cu/CNT nanocomposite fabrication with different raw material properties using a planetary ball milling process. Powder Technol 323:563–573
25. Dash P, Dash T, Rout TK, Sahu AK, Biswal SK, Mishra BK (2016) Preparation of graphene oxide by dry planetary ball milling process from natural graphite. RSC Adv 6:12657

Effect of Road Conditions on Human Subjects in Sitting and Standing Posture Using FEM

Shubham Sharma, Jagjit Singh, Sachin Kalsi, Ishbir Singh, and Manjot Kaur

Abstract Human subjects are exposed to vibration while performing different activities in both sitting and standing postures. The amplitude and intensity of vibration vary according to road conditions, environmental conditions, time and other related factors. It becomes necessary to find out the effect of vibration on human subjects in terms of transmissibility, natural frequencies and resonant frequency. In the present study, a 4-layer CAD model of a human subject has been used to perform a harmonic analysis, i.e., applying a sinusoidal input with an acceleration of 1 m/s^2 and a frequency range of 0–20 Hz in the vertical direction. The CAD model of the human subject has been developed using 95th percentile anthropometric data and consists of four different anatomy layers, i.e., organs, skin, muscles and bones. The feet-to-head transmissibility has been evaluated for each posture and compared with the available experimental results in the literature. The transmissibility in sitting posture was found to be more in comparison to standing posture.

Keywords Biodynamics · Biomechanics · FEM · Human factors · Road conditions · Transmissibility

Nomenclature

CAD	Computer-Aided Design
Hz	Hertz
FEM	Finite Element Method

S. Sharma · J. Singh · S. Kalsi (✉)
Department of Mechanical Engineering, Chandigarh University, Mohali, India
e-mail: phd.sachinkalsi@gmail.com

I. Singh
Department of Mechanical Engineering, Indus University, Ahmedabad, India

M. Kaur
Lead Engineer-Testing, Fidelity Information Services Ltd., Mohali, India

A. Sachdeva et al. (eds.), *Recent Advances in Operations Management and Optimization*, Lecture Notes in Mechanical Engineering,
https://doi.org/10.1007/978-981-99-7445-0_3

1 Introduction

Human subjects travelling by car, railway train or bus in sitting or standing posture for long hours may lead to unbearable pain in different body parts due to the vibrations induced by varying road conditions, suspension, high speed, etc. of the vehicle. These vibrations when exceeding the prescribed limits can cause permanent damage to the human subject. It is important to find out the vibrational effect on the human in standing as well as sitting posture when exposed to varying road conditions. Biodynamic models representing different anthropometric data have been used to find out the different vibration parameters that include human comfort [1], transmissibility [2], frequency range, resonance frequency [3, 4], apparent mass [5], phase angle [6], etc. The biodynamic models include multi-body dynamic models (MBD) [7], LPM [8, 9] and FEM [10–13]. In a few of the studies, several human subjects have been considered to experimentally find out the vibration parameters under different operational conditions like agriculture, road conditions (rough, smooth), etc. [14–17]. With the variation in the type of vibration, load input, modal analysis, harmonic and random analysis have been conducted in the existing literature. The human subject posture has been considered to be either standing [6–8, 18–20] or sitting [21–23] and the direction of vibration load has been applied in the vertical, fore-aft and lateral direction [18, 21, 24–26]. The impact of vibration on different parts has also been evaluated in the existing literature [13, 27].

In the present study, two different postures of the human subject, i.e., sitting and standing have been considered to evaluate the transmissibility of the human subject for sinusoidal external vibrations at an acceleration of 1 m/s^2 and the range of frequency is 0–20 Hz.

2 Methodology

In this study, the 95th percentile of Indian anthropometric data [28] (covering the largest population) corresponding to 76 kg human subjects in a standing and sitting position was considered to create a 4-layer 3D representation of the CAD model of the body. Humans develop and test the subject using SOLIDWORKS 2018, which includes four different anatomical layers of the human body, i.e., organs, skin, muscles and bones. The biomechanical properties of organs, skin, muscles and bones of the human subject have been taken from the existing literature [29]. Boundary conditions used while performing a simulation study are as follows:

- A CAD model has been considered in standing and sitting posture (seated on the seat cushion) while acceleration applied is in the vertical direction.
- Both feet are in direct contact with the floor and fixed in case of both standing as well as sitting posture, where the buttocks are always in contact with the seat cushion in the sitting position as shown in Fig. 1a, b.

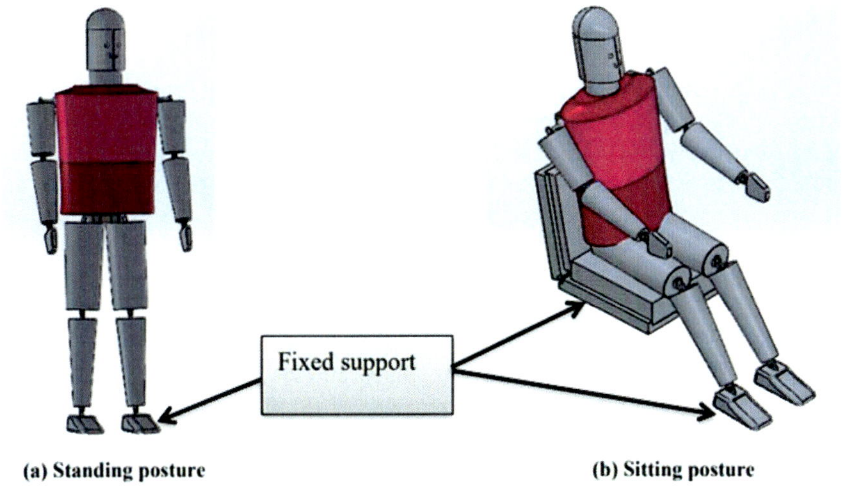

Fig. 1 Boundary conditions

- STHT and FTHT transmissibility has been evaluated in sitting and standing posture at 1 m/s^2 excitation amplitude applied in the vertical direction.
- The range of frequency 0–20 Hz has been considered for the study as the human body experiences bad impacts like movement infection and spinal pain that is mostly fixed support found in the human subjects while travelling and driving vehicles in the considered frequency range [30].
- No outer load or force has been applied to the human body.

Harmonic response analysis has been performed on the developed CAD model of a human subject using ANSYS 18.2. Individual parts of the human subject have been meshed using tetrahedral and hexahedral mesh elements for 3D parts and for 2D, quad and tri have been used. The reason to use tetrahedral mesh elements is due to the complex structure of the human subject. STHT and FTHT are the ratio that signifies the range to which the input vibrations have been transmitted to the different parts of the body. A particular road condition has been considered that represents an acceleration of 1 m/s^2 and a frequency range of 0–20 Hz that has been applied to the driver of the vehicle.

3 Results and Discussions

The FTHT and STHT for a human subject in standing and sitting posture have been evaluated under fixed road conditions, i.e., acceleration of 1 m/s^2 and the range of frequency is 0–20 Hz, as discussed below.

3.1 Standing Posture

The FTHT was evaluated on 76 kg human subjects using 4-layered CAD models (bone, organ, muscle and skin) developed by considering 95th percentile anthropometric data to determine the effect of vibration and a frequency range that leads to more harm to human subjects by determining the transmissibility from feet to the head while considering the acceleration magnitude as 1 m/s^2 [25] and frequency range of 0–20 Hz in the vertical direction. Figure 2 indicates the ratio or transferability of the acceleration applied to the legs to the acceleration received from the head.

According to Fig. 2, during a study on a 4-layer CAD model (bones, organs, muscles, and skin) the value of transmissibility at 6 Hz was found to be 1.44 and decreases further with the increase in frequency. The gradual decrease in transmissibility value just after the peak point could be due to the human subject becoming more resistant to an external excitation frequency. Figure 2 shows that as frequency increases beyond 6–8 Hz, the transmission value decreases. This may be due to the different layers of the human body because the bones are hard, e.g., rigid in nature, and have fewer damping properties, whereas organs, muscles and skin are soft tissues with a higher damping value. Because of their complex properties, muscles and tissues in the human body can dampen vibrations [29].

Transmissibility of the current study's 4-layer CAD model has been validated with a study by Garg et al. [14] that measures human vibration transmissibility in the range of frequency is 1–50 Hz with two different damped vibration effects using an LPM model. According to Fig. 2, in the study conducted by Garg et al. [14], the first maximum transmissibility peak, with a value of 1.74, occurs at 6 Hz, most likely because the upper body depends on the spine's rigidity. A possible explanation for the second-highest peak, 1.63 at 19.4 Hz, which is closer to 20 Hz, is that the mass of the entire body rests on the rigidity of the legs. For the transmission rate study, two peaks

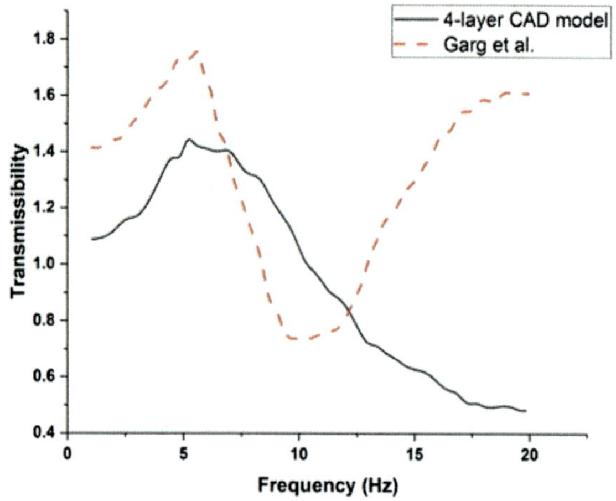

Fig. 2 Comparison of Transmissibility obtained for standing posture in the current study and existing literature

were discovered because of two different damping rates of 0.422 for frequencies 10–20 Hz and 0.37 for frequencies 0–10 [14]. When the results of a 4-layer CAD model were compared to the results of the current study, it was discovered that there were differences between the outcomes. This could be because different biomechanical characteristics and each segment of a human subject have been considered closer to the average value for the entire body.

It was discovered that the transmissibility values were closer to each other up to frequency 0–10 Hz, with a damping ratio of 0.377. Then there was a difference because the current study used the same damping ratio of 0.377 from 11 to 20 Hz while the validated study used a different damping ratio of 0.422 from 11 to 20 Hz. It was also discovered the maximum transmissibility effect occurs at 6 Hz, considered excitation magnitude will cause discomfort in the human body.

3.2 *Sitting Posture*

The STHT measurement for a 4-layer CAD model in a sitting position at an angle of 0° as a backrest angle at an acceleration of 1 m/s^2 and the range of frequency 0–20 Hz is shown in Fig. 3. The first effect is observed at a 6 Hz frequency with a value of transmissibility of 3.72 and further, the value of transmissibility decreases up to 8 Hz which results in the discomfort feeling in the human subject. The second peak of transmissibility with the value of 1.32 is observed at a frequency of 12 Hz and starts decreasing continuously as shown in Fig. 3. After this, the value of transmissibility decreases, which may be due to the backrest support at an angle of 0° inclination.

For the validation of results obtained in sitting posture, the study conducted by Wang et al. [31] and Cho and Yoon [32] has been considered as shown in Fig. 3. Wang et al. [31] conducted a study on 12 male subjects in multiple postures: without, vertical and inclined backrest support in vertical and fore-and-aft direction at excitation rates of 0.25, 0.5 and 1.0 m/s^2 between 0 and 15 Hz. The vibration effect was evaluated by taking an excitation acceleration of 1 m/s^2 in the vertical direction, the greatest vibration effect occurred at 5 Hz of 1.78 and the second at 10 Hz of 1.

Cho and Yoon [32]used harmonic analysis was used to analyze automotive ride quality in a biomechanical model with 1, 2, 3 and 9° of freedom in a seated position with the back supported at various tilt angles. The effect of vibration on the hip, back, and head of 10 subjects was measured by vertical floor vibration and 3 transmission effects were obtained for each subject. The results for the model with 9 DOF correspond to good vibrational effects, with the maximum effect at 4.2 Hz of 3.2223 and a second effect at 7.7 Hz of 1.00. When these results were compared to the validated results in seating posture, it was discovered that they were closer to the validated results.

The current study discovered that when human subjects were exposed to vibration while standing and sitting, the organs (heart, lungs, stomach and liver) were more affected due to strong accelerations in the frequency range of 4–8 Hz. Analysis revealed that vibration has the greatest impact on the head, causing headaches,

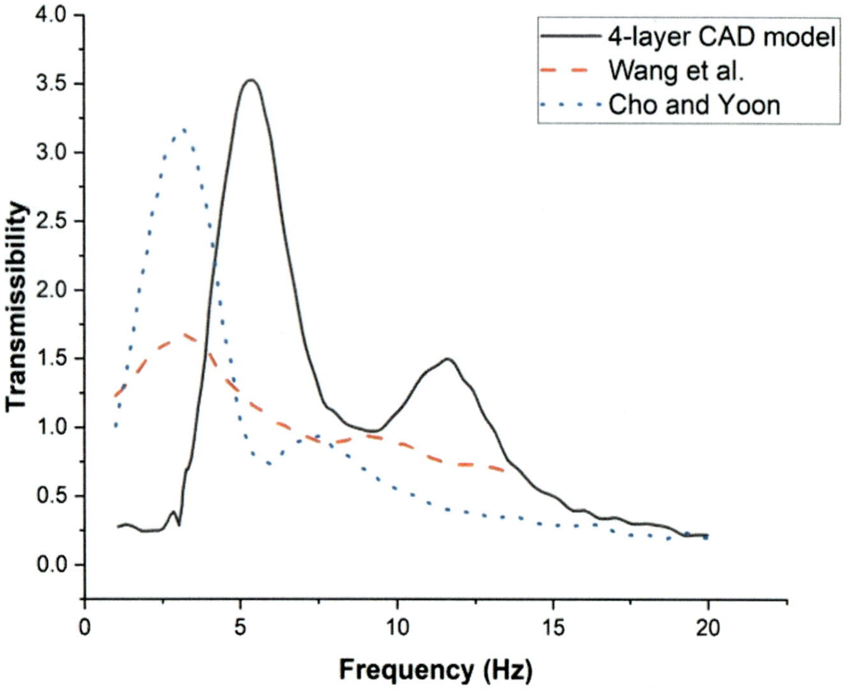

Fig. 3 Comparison of transmissibility in sitting posture of the current study and the existing literature

motion sickness and other symptoms. During the analysis, the lower arm of the human subjects was affected the most, resulting in pain and sensation in the hand. According to the results of this study, human subjects will experience discomfort at the considered excitation magnitude at a frequency of 6 Hz due to the maximum transmissibility effect.

4 Conclusions

In the current study, a 4-layered concept was used to develop a CAD model of a human subject that includes bones, organs, muscles and skin so that variations in biodynamic properties of the human subject can be accommodated while performing a study to observe the response of the human subject using the FEM approach. A study was conducted to determine the effect of vibration by measuring the transmissibility from FTHT in a standing posture and STHT in a sitting posture while exposed to vertical vibration at an acceleration of 1 m/s^2 in the 0–20 Hz frequency range. In sitting posture, the first effect is observed at a 6 Hz frequency with a value of transmissibility as 3.72 and further, the value of transmissibility decreases up to 8 Hz, which results in discomfort in the human subject. The second peak of transmissibility with the value of 1.32 is observed at a frequency of 12 Hz and starts decreasing continuously. In standing posture, the value of transmissibility at 6 Hz was found to be 1.44 and decreased further with the increase in frequency, the value of transmissibility at 6 Hz was found to be 1.44 and decreased further with the increase in frequency. The standing and sitting posture results were compared and validated against the existing literature.

References

1. Goggins K, Godwin A, Lariviere C, Eger T (2016) Study of the biodynamic response of the foot to vibration exposure. Occup Ergon 13:53–66. https://doi.org/10.3233/OER-160236
2. Mansfield NJ, Griffin MJ (2000) Non-linearities in apparent mass and transmissibility during exposure to whole-body vertical vibration. J Biomech 33:933–941. https://doi.org/10.1016/S0021-9290(00)00052-X
3. Gupta M, Gupta TC (2017) Modal damping ratio and optimal elastic moduli of human body segments for anthropometric vibratory model of standing subjects. J Biomech Eng 139. https://doi.org/10.1115/1.4037403
4. Randall JM, Matthews RT, Stiles MA (1997) Resonant frequencies of standing humans. Ergonomics 40:879–886. https://doi.org/10.1080/001401397187711
5. Amirouche FML, Ider SK (1988) Simulation and analysis of a biodynamic human model subjected to low accelerations-a correlation study. J Sound Vib 123:281–292. https://doi.org/10.1016/S0022-460X(88)80111-1
6. Matsumoto Y, Griffin MJ (2011) The horizontal apparent mass of the standing human body. J Sound Vib 330:3284–3297. https://doi.org/10.1016/j.jsv.2011.01.030
7. Fritz M (2000) Simulating the response of a standing operator to vibration stress using a biomechanical model. J Biomech 33:795–802. https://doi.org/10.1016/S0021-9290(00)00038-5
8. Matsumoto Y, Griffin MJ (2003) Mathematical models for the apparent masses of standing subjects exposed to vertical whole-body vibration. J Sound Vib 260:431–451. https://doi.org/10.1016/S0022-460X(02)00941-0
9. Singh I, Nigam SP, Saran VH (2015) Modal analysis of human body vibration model for Indian subjects under sitting posture. Ergonomics 58:1117–1132. https://doi.org/10.1080/00140139.2014.961567
10. Kumar R, Singh I, Kalsi S, Sehgal SS (2018) Modal analysis of standing human body using FEM. J Sci Ind Res (India) 77:639–645
11. Kitazaki S, Griffin MJ (1997) A modal analysis of whole-body vertical vibration, using a finite element model of the human body. J Sound Vib 200:83–103. https://doi.org/10.1006/jsvi.1996.0674
12. Privitzer E, Belytschko T (1980) Impedance of a three-dimensional head-spine model. Math Model 1:189–209. https://doi.org/10.1016/0270-0255(80)90037-8
13. Bartz JA, Gianotti CR (1975) A computer program to generate dimensional and inertial properties of the human body. J Manuf Sci Eng Trans ASME 97:49–57. https://doi.org/10.1115/1.3438590
14. Ross MA (1976) Vertical mode human body vibration transmissibility. IEEE Trans Syst Man Cybern SMC-6 102–112. https://doi.org/10.1109/TSMC.1976.5409180
15. Baker WDR, Mansfield NJ (2010) Effects of horizontal whole-body vibration and standing posture on activity interference. Ergonomics 53:365–374. https://doi.org/10.1080/00140130903402242

16. Subashi GHMJ, Matsumoto Y, Griffin MJ (2006) Apparent mass and the cross-axis apparent mass of standing subjects during exposure to vertical whole-body vibration. J Sound Vib 293:78–95. https://doi.org/10.1016/j.jsv.2005.09.007
17. Parsons KC, Griffin MJ (1988) Whole-body vibration perception thresholds. J Sound Vib 121:237–258. https://doi.org/10.1016/S0022-460X(88)80027-0
18. Thuong O, Griffin MJ (2011) The vibration discomfort of standing persons: 0.516-Hz fore-and-aft, lateral, and vertical vibration. J Sound Vib 330:816–826. https://doi.org/10.1016/j.jsv.2010.08.040
19. Zheng X, Brownjohn JMW (2001) Modeling and simulation of the human-floor system under vertical vibration. In: Proceedings of SPIE—the international society for optical engineering, pp 513–520
20. Matsumoto Y, Griffin MJ (1998) Dynamic response of the standing human body exposed to vertical vibration: influence of posture and vibration magnitude. J Sound Vib 212:85–107. https://doi.org/10.1006/jsvi.1997.1376
21. Thuong O, Griffin MJ (2012) The vibration discomfort of standing people: relative importance of fore-and-aft, lateral, and vertical vibration. Appl Ergon 43:902–908. https://doi.org/10.1016/j.apergo.2011.12.011
22. Matsumoto Y, Griffin MJ (2000) Comparison of biodynamic responses in standing and seated human bodies. J Sound Vib 238:691–704. https://doi.org/10.1006/jsvi.2000.3133
23. Paddan GS, Griffin MJ (1993) The transmission of translational floor vibration to the heads of standing subjects. J Sound Vib 160:503–521. https://doi.org/10.1006/jsvi.1993.1041
24. Subashi GHMJ, Matsumoto Y, Griffin MJ (2008) Modelling resonances of the standing body exposed to vertical whole-body vibration: effects of posture. J Sound Vib 317:400–418. https://doi.org/10.1016/j.jsv.2008.03.019
25. Harsha SP, Desta M, Prashanth AS, Saran VH (2014) Measurement and bio-dynamic model development of seated human subjects exposed to low-frequency vibration environment. Int J Veh Noise Vib 10:1–24. https://doi.org/10.1504/IJVNV.2014.059627
26. Tarabini M, Solbiati S, Moschioni G, Saggin B, Scaccabarozzi D (2014) Analysis of the non-linear response of the human body to vertical whole-body vibration. Ergonomics 57:1711–1723. https://doi.org/10.1080/00140139.2014.945494
27. Raines JK, Jaffrin MY, Shapiro AH (1974) A computer simulation of arterial dynamics in the human leg. J Biomech 7:77–91. https://doi.org/10.1016/0021-9290(74)90072-4
28. Indian anthropometric dimensions for ergonomic design practice—Debkumar Chakrabarti—Google Books. https://books.google.co.in/books/about/Indian_anthropometric_dimensions_for_erg.html?id=koyAAAAAMAAJ&redir_esc=y
29. Bhatia A, Kalsi S, Sehgal AK, Singh I (2022) Comparative study of different seat cushion materials to improve the comfort of tractor seat. J Inst Eng Ser A 103(2):387–396. https://doi.org/10.1007/S40030-022-00622-8
30. Singh A, Singh I, Kalsi S (2020) Transmissibility evaluation of whole-body vibration using three-layer human CAD model. J Inst. Eng Ser C 101:595–602. https://doi.org/10.1007/s40032-020-00559-6
31. Wang W, Bazrgari B, Shirazi-Adl A, Rakheja S, Boileau P (2010) Biodynamic response and spinal load estimation of the seated body in vibration using finite element modelling. Ind Health 48:557–564. https://doi.org/10.2486/indhealth.MSWBVI-34
32. Cho Y, Yoon Y-S (2001) Biomechanical model of human on the seat with backrest for evaluating ride quality. Int J Ind Ergon 27:331–345. https://doi.org/10.1016/S0169-8141(00)00061-5

Design and Development of High-Precision Scanning Flexural Mechanism Using PID

Shrishail Sollapur, Tarang Shinde, Satish Raut, Abhijit Atpadkar, Prashant Nimbalkar, and Mahesh Rathod

Abstract Rigid linkage mechanism has narrow accuracy and repeatability. The mechanism offers motion between the joints. A different approach has been made to develop S-type flexure mechanism for high-precision systems. This paper presents the flexural mechanism design with the building blocks. This S-shaped mechanism is developed using a numerical approach to get linear motion. The mechanism is incorporated with a voice coil motor (VCM), optical encoder, and dSPACE 1104 R&D microcontroller board. The S-shaped mechanism is integrated, designed, and developed with system integration and identification of the system. The proportional–internal–derivative (PID) controls the implementation of the mechanism. The system identification is done and performance is evaluated for the natural frequency, damping factor and also evaluated by experimentation for the same. The transfer function is used to build the control system and parameters are tuned with the help of PID. PID implementation in real time imparts the accuracy for positioning at a high scanning speed of 0.5 mm/s.

Keywords Flexure mechanism · PID control · DFM · VCM · dSPACE1104

S. Sollapur (✉)
Faculty of Engineering and Technology, Department of International Institute for Aerospace Engineering and Management (IIAEM), Jain (Deemed-to-be University), Bengaluru, Karnataka, India
e-mail: shrishail.sollapur@gmail.com

T. Shinde
Faculty of Engineering, Department of Mechanical Engineering, Yashoda Technical Campus (Affiliated to Dr. Babasaheb Ambedkar Technological University, Lonere, District: Raigad), Satara, Maharashtra, India

S. Raut · A. Atpadkar · P. Nimbalkar · M. Rathod
Faculty of Engineering, Department of Mechanical Engineering, Yashoda Technical Campus (Affiliated to Dr. Babasaheb Ambedkar Technological University, Lonere, District: Raigad), Satara, Maharashtra, India

A. Sachdeva et al. (eds.), *Recent Advances in Operations Management and Optimization*, Lecture Notes in Mechanical Engineering,
https://doi.org/10.1007/978-981-99-7445-0_4

1 Introduction

The advances in the area of manufacturing, electronics, and materials have increased the precision technology in the present scenario [1, 3]. Precision manipulators with higher economic and performance features have enhanced small-scale technologies. In modern technology, nano- and micro-positioning stages are essential [1, 4]. Their usages are in applications like micromachining, confocal microscope, and scanning probe. A variety of XY mechanisms are developed in the wide area of the screw-type precision ball recirculation mechanisms. The previously developed XY mechanism has restrictions like limited range performance characteristics and accuracy. The control system is developed to obtain the required performance. A new method to build mechanisms such as flexural mechanisms with high-speed precision applications was developed [5–7]. Flexures are developed on elasticity of material for operation [8–10]. Motions are developed because of anamorphisms at the molecule level with two characteristics accuracy and high-speed applications. The advantage of the flexure mechanism is the smooth motion without friction and backlash. The flexure mechanisms are simple in construction (monolithic) and assembly. They have repeatability and also are predictable [11–13]. This work attempts to build and analyze motion with a large motion range. The degrees of freedom are obtained by the flexure mechanism. These are grouped into two types beams and hinges. The hinges are used for rotational motion and beams are used for planar motions.

2 Flexure Mechanism

Flexure mechanism provides mobility between the stage and fixed support. The motion is given by the interfacing element as explained in Fig. 1.

The DFM gives consideration merit over the conventional motion in characteristics as no friction loss and no lubrication is required. The interface element is replaced with beams eliminating friction, backlash, and high repeatability. The S-shaped flexure mechanism is shown in Fig. 2.

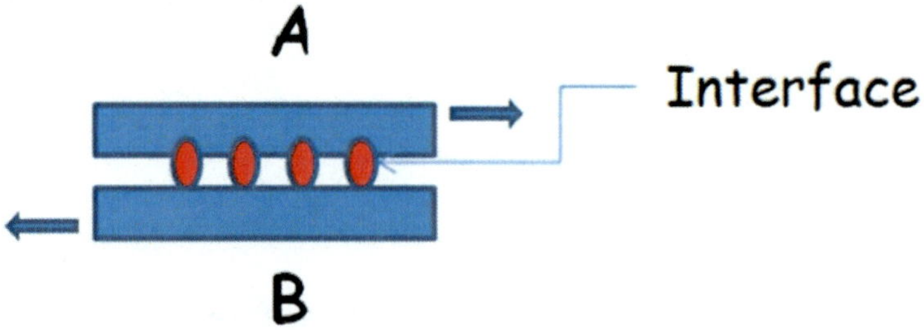

Fig. 1 Interfacing element [2]

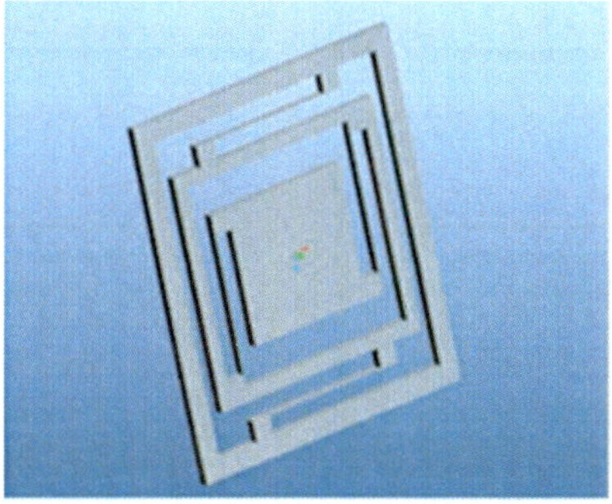

Fig. 2 S-shaped flexure mechanism

3 Analysis of Flexure Mechanism

The analysis of the flexure mechanism indicates that it proves null parasitic error and linear motion.

The deflection of S-shaped mechanism is given by Eq. 1.

$$\delta = \frac{FL^3}{12EI} \tag{1}$$

where

L = Flexure beam length in mm

F = Applied force in N

E = Beam material elastic modulus in N/mm^2

I = Mass moment of inertia mm^4

Rotation is given in Eq. 2,

$$\theta = t^2\left(\frac{1}{b1^2} + \frac{1}{b2^2}\right) \times \frac{\delta}{L} \tag{2}$$

where

t = Beam thickness in mm

b = Beam width in mm

Parasitic Error,

The load is applied to the mechanism of 1 N which produces a deflection of 0.4048 mm as shown in Fig. 3. The stresses are well within the range as shown in Fig. 4

The flexure mechanism is built and developed using S-shaped double flexure mechanism. The experimental setup is shown in Fig. 5.

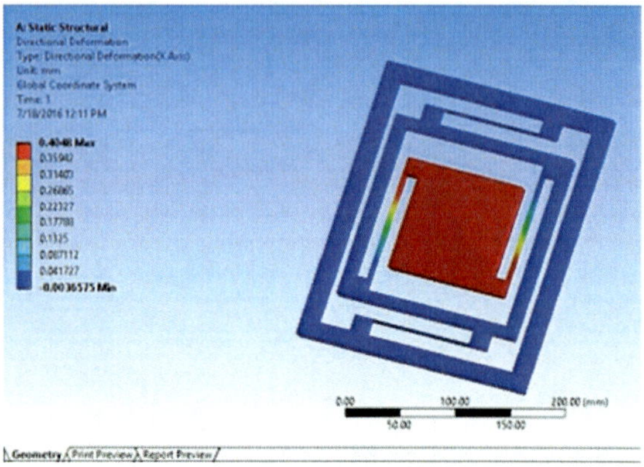

Fig. 3 Deflection of flexure mechanism

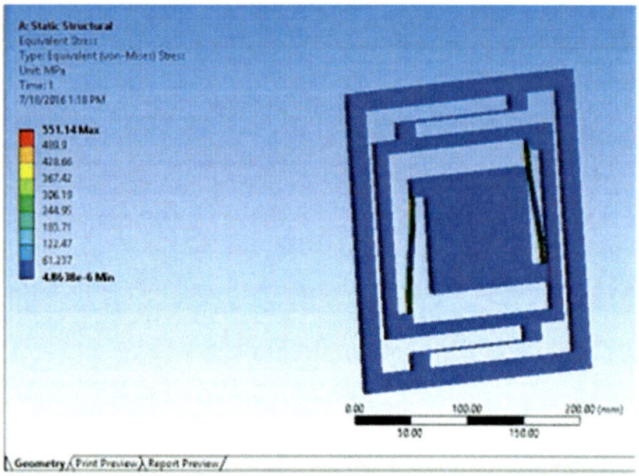

Fig. 4 Stress developed in flexure mechanism

Fig. 5 Actual model of flexure mechanism

Fig. 6 Experimental setup

4 Mechatronics Integration of the System

The flexure mechanism is integrated with dSPACE DS 1104 microcontroller and actuator. The system is actuated by a voice coil motor (VCM) with applied force converting to current and voltage. The given current is low; therefore, it is amplified using LCAM. Simulink MATLAB file is prepared to compare the signal obtained from the encoder with a reference signal to evaluate the error signal. The motion head is moved by a given force. The encoder is mounted in the system and motion is detected. The values gained from the optical encoder are sent to dSPACE and saved in a mat file.

The overall mechatronics integration setup is shown in Fig. 6.

The different parameters like stiffness, damping factor are evaluated with dSPACE DS1104. The system natural is calculated from the setup.

5 Stiffness of the System

Stiffness is determined by the experimental force–deflection curve. The MATLAB code is built to control and actuate the voice coil motor. The position of the motion head is read by optical encode. The stiffness is evaluated for to and fro motion and obtained as shown in Fig. 7.

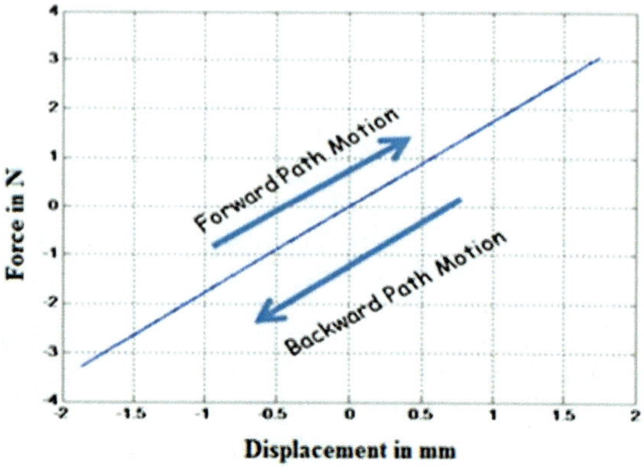

Fig. 7 Force–deflection curve

6 Damping Factor

As the initial state abruptly shifts from zero to one in relation to time, the system's temporal behaviour is defined. The amplitude value is set to zero to determine how the system responds to abrupt input for substantial deviations over a longer period of time. The output must stabilize at a steady state in order to get the steady state of the system's step response.

Experimental results are plotted on the graph (Fig. 8). From Fig. 8, both the damping factor and damped natural frequency are calculated. The motion head is given the free to and fro movement, finally coming to rest position.

The damping factor is evaluated from the logarithmic decrement and is given by

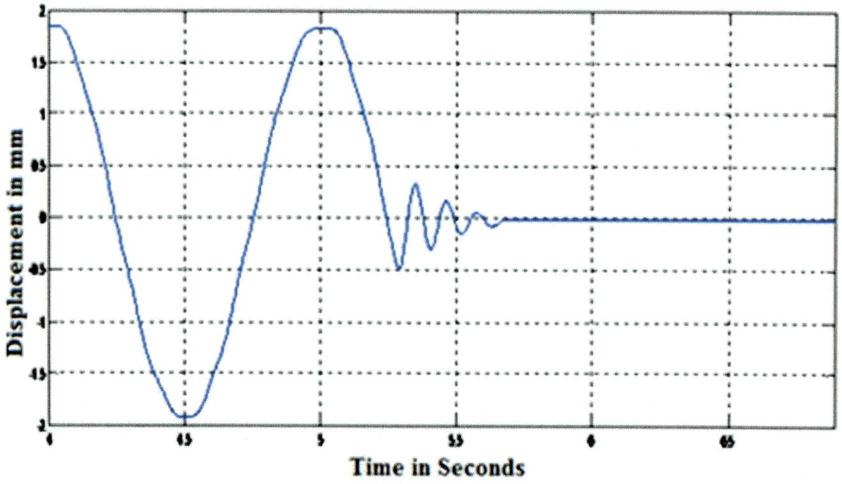

Fig. 8 The plot of displacement versus time

$$\delta = \frac{1}{n}\left[\log\left(\frac{X_0}{X_n}\right)\right]. \quad (3)$$

where

X_0 = First peak amplitude

X_n = Amplitude to peak at n periods

n = Successive number of peaks
Damping factor

$$\xi = \frac{\delta}{\sqrt{4\pi^2 - \delta^2}} \quad (4)$$

From the experimentation, the obtained values are

$$\delta = 1.1972, \xi = 0.19402$$

7 Identification of the System

An identification system is required to build the transfer function of the S-shaped flexure mechanism, to control the input signal to the actuator and positioning of the motion head. MATLAB code is built for the estimation of frequency response with an amplitude of 1 mm and is given a frequency response graph for the system to build. The peak frequency of 13.52 rad/s is obtained. The transfer function is developed using the experimental data of the system. The transfer function of the present model is

$$G(S) = \frac{1}{0.00963s^2 + 0.05055s + 1.762} \quad (5)$$

8 PID Implementation

The PID system feedback controller is adopted in industrial applications as shown in Fig. 9. The change between the variable being measured and the set point is evaluated by the PID controller using an error value. The PID controller uses influence variables to modify the process in an effort to decrease error. The PID controller is composed of three parameters the proportional, integral and derivatives values represented by P, I, D.

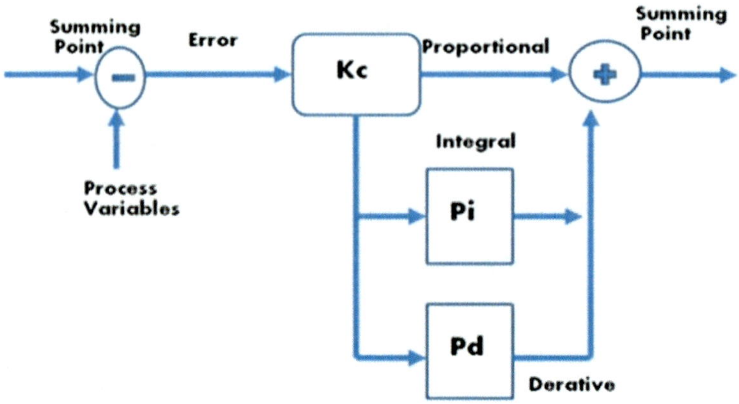

Fig. 9 PID controller

The proportional integral derivative controller is tuned using the Ziegler–Nichols approach. John G. Ziegler and Nathaniel B. Nichols constructed it. The system is implemented by setting the derivative (D) and integral (I) values to zero. The proportional increases as (P) from zero until it approaches the ultimate gain ('U') and oscillates with the same amplitude. The controller being utilized will determine the ultimate gain (U) and time cycle (T) values, which are used to construct the P, I, and D gains [7–9].

From Fig. 10, PID control result in real time on S-shaped flexure mechanism at lower velocity (0.5 Hz frequency with Speed = 300 μm/s) and lesser range of scan (Amplitude = 260 μm) for motion stage has been indicated. The accuracy of displacement is less than 0.5 μm.

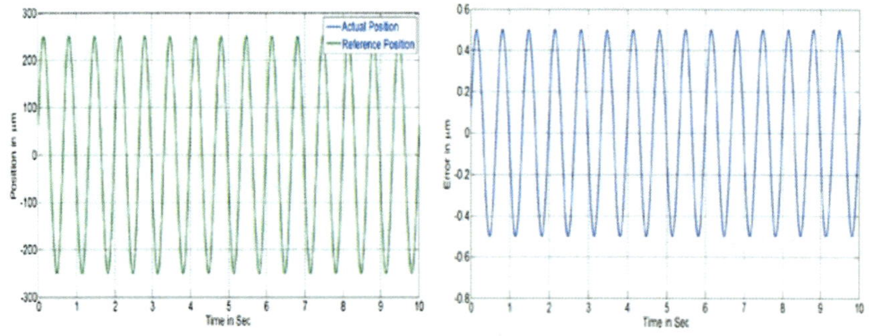

(a): Comparing actual position with required reference (b): Real time positioning error positioning

Fig. 10 PID implementation on S-shaped flexure mechanism at 260 μm amplitude with 0.5 Hz frequency

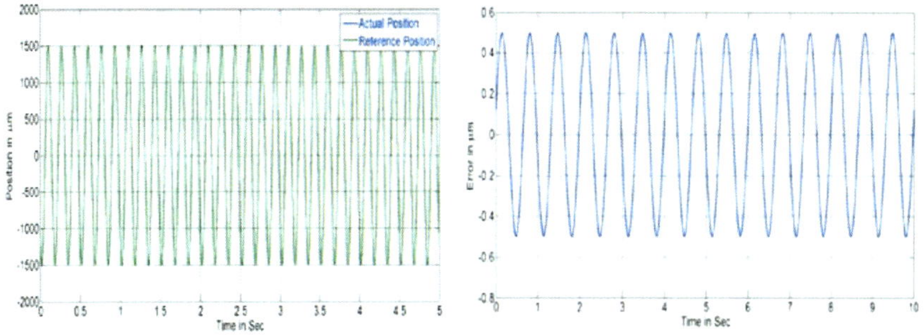

(a): Comparing actual position with required reference (b): Real time positioning error

positioning

Fig. 11 PID implementation on S-shaped flexure mechanism at 1500 μm amplitude and 3 Hz frequency

PID control results in real time on S-Shaped flexure mechanism are comparatively higher speed (3 Hz frequency with speed = 1000 μm/s) and range of scan (Amplitude = 1500 μm) of motion stage (Fig. 11). The accuracy of deflection is less than 3 μm.

9 Conclusion

The S-shaped flexure mechanism is designed for precision applications. It is experimentally built and developed with dSPACE DS1104 microcontroller. Both static and dynamic parameters are characterized by experimental and theoretical agreed. PID control is implemented on S-shaped flexure mechanism using the dSPACE DS1104 Control desk environment. 0.5 μm of precision is reached at a slower scanning speed thanks to the Zeigler–Nicholas tuning algorithm being used to adjust the PID settings. The positioning accuracy of 3 μm is obtained at a higher speed of scanning. The least value of error and rejection is being obtained to track at various frequencies. For the regulation, LQR control can be used and LQI control accomplishes good distribution during rejection. The LQI method can adopt to track feed-forward control with other strategies on setup.

References

1. Qin Y, Shirinzadeh B, Zhang D, Tian Y (2013) Design and kinematics modeling of a novel 3-dof monolithic manipulator featuring improved Scott-Russell mechanisms. J Mech Des 135(10):101004. https://doi.org/10.1115/1.4024979
2. Awatar S (2004) Synthesis and analysis of parallel kinematic XY flexure mechanism, PhD Thesis, MIT, Cambridge, MA

3. Deshmukh S, Gandhi PS (2003) Optomechanical scanning systems for microstereolithography (MSL): analysis and experimental verification. J Mater Process Technol 209(3):1275–1285
4. Awtar S, Slocum AH (2005) Design of parallel kinematic XY flexural mechanism. In: Proceedings of IDETC/CIE international design engineering technical conferences & computers and information in engineering, 2005, California, USA
5. Lescano S, Rakotondrabe M, Andreff N (2012) Micromechanisms for laser phonosurgery: a review of actuators and compliants parts. In: International conference on biomedical robotics and biomechatronics, Rome, Italy
6. Ramesh T, Bharanidaran R, Gopal V (2014) Design and development of XY micro-positioning stage using modified topology optimization technique. App Mech Mat 592–594:2220–2224
7. Deshmukh S, Gandhi PS (2007) A novel optomechatronic focused laser spot submicron scanning system for microstereolithography. In: Poster presentation at Bangalore Nano conference 2007. International conference on nanotechnology
8. Zhu WL, Zhu Z, Shi Y, Wang X, Guan K, Ju BF (2016) Design, modeling, analysis and testing of a novel piezo-actuated Xy compliant mechanism for large workspace nano-positioning. Smart Mater Struct 25(11):115033
9. Kim HS, Cho YM (2009) Design and modeling of a novel 3-DOF precision micro-stage. Mechatronics 19:598–608
10. Sollpaur S, Patil M, Deshmukh S (2018) Evaluation of stiffness and parametric modeling of XY flexure mechanism for precision applications. J Model Simul Mater 1:8–15
11. Li Y, Xu Q (2009) Modeling and performance evaluation of a flexure-based XY parallel micromanipulator. Mech Mach Theory 44:2127–2152
12. Sollapur S, Patil MS, Deshmukh SP (2018) Design and development aspects of flexure mechanism for high precision application. In: AIP conference proceedings 1943, p 020023. https://doi.org/10.1063/1.5029599
13. Liu Y-T, Li B-J (2010) Precision positioning device using the combined piezo-VCM actuator with frictional constraint. Precis Eng 34:534–545

Identification of Constraint in Healthcare Unit by Using Dice Game

Mohit Datt, Ajay Gupta, and Sushendra Kumar Misra

Abstract Health care is the world's largest and fastest-growing sector- both per revenue as well as employment. Many operation management techniques have been implemented in the healthcare sector to improve the utilization of resources and the performance of systems. Theory of constraints was extensively implemented and resulted in positive outcomes in the healthcare sector in the literature. Identification of constraints is very crucial for system improvements and a dice game was introduced and also witnessed a number of applications in different environments. But for the healthcare industry, it is still novel. In this paper, we tried to identify the constraints in the complex and uncertain environment of the healthcare sector. Hence, this paper will guide working academicians and professionals to understand the concept of the dice game for better improvement of healthcare units.

Keywords Theory of constraints · Dice game · Healthcare · Operations management

1 Introduction

In the 1970s and 1980s, many efforts were devoted to balancing line manufacturing and the Theory of Constraints (TOC) showed concepts that discussed the roadmap for balancing production lines perfectly. TOC was introduced by Dr. Eliyahu Goldratt to identify the bottlenecks because constraints always prevent the system from achieving its goals. TOC philosophy is a combination of many approaches as well as methods like five focusing steps, throughput accounting, logical thinking process trees, critical chain project management, cause & effect, root causes & undesirable

M. Datt (✉) · A. Gupta
Department of Industrial and Production Engineering, Dr B R Ambedkar National Institute of Technology, Jalandhar 144027, India
e-mail: mohitdatt1991@gmail.com

S. K. Misra
I.K. Gujral Punjab Technical University, Jalandhar, India

A. Sachdeva et al. (eds.), *Recent Advances in Operations Management and Optimization*, Lecture Notes in Mechanical Engineering,
https://doi.org/10.1007/978-981-99-7445-0_5

effects, drum-buffer-rope, and buffer management for continuous improvement of the system [1–4]. TOC has been tested for its potential and importance in different fields throughout the globe by many researchers. Rich scholarly literature witnessed many domains such as operations management [5, 6], banking [7], product-mix decision [8, 9], aerospace sector [10], supply chain [7], project management [11], service sector [12, 13], manufacturing [14, 15], health care [16, 17], and many more.

In the healthcare sector, providing error-free treatment, the effectiveness of processes, proper utilization of critical resources, and enhanced patient care are the major goals. TOC can be an effective philosophy in delivering more patients and providing services with limited resources in the complex and uncertain environment of the healthcare sector. In the novel "*The Goal*", the author discusses his thoughts on running manufacturing units smoothly and also introduces accounting parameters. Several researchers tried to simulate and measure throughput, inventory, and operating expenses of different working environments with a "dice game" [18–20]. Here in this study, we tried to introduce this concept for demonstration in the uncertain and complex balanced healthcare system. Due to statistical fluctuations and dependent processes balanced systems are difficult to manage.

The main objective of this study was to guide practitioners and academics about concepts of dice in healthcare and provide knowledge to:

- Demonstrate the complex healthcare system under statistical fluctuations and improve the flow of patients
- Improve the flow of patients using TOC concepts.

2 Procedure for Dice Game

ABC healthcare unit provides outpatient services to improve the mental health of patients. The unit treated patients with experienced staff and consisted of seven departments including registration, nursing care, psychosocial evaluation, psychiatric evaluation, medication, therapy, and termination. Figure 1 represents the flow of patients in this unit.

2.1 Flow of Patients

This healthcare unit works as a push system, where the patient is moved next department after the diagnosis in the previous department. Firstly, the patient reports at the registration desk and then moves forward for the triage in the nursing care unit. The nursing care unit diagnoses the patient and if psychiatric evaluation is required then the patient moves toward the psychiatric unit otherwise patient will forward toward the psychosocial evaluation. After the psychiatric unit medication visit is required for the patient. The psychosocial evaluation will forward the patient to the therapy department unless the patient requires a psychiatric evaluation. At last, the patient

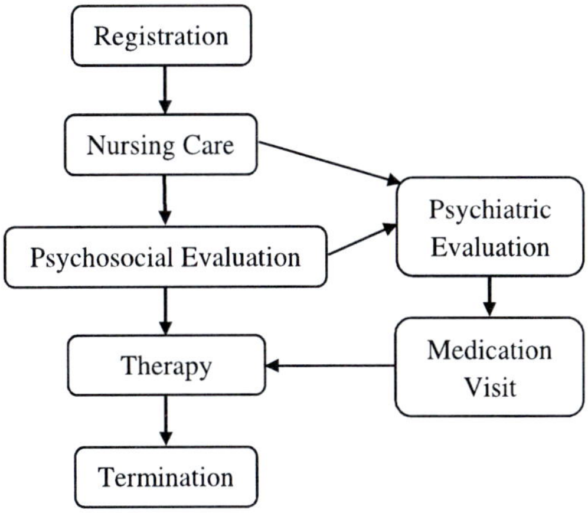

Fig. 1 Layout of the healthcare unit

visits the termination department for all the documentation and further suggestions about the medicines and precautions.

2.2 *Rules for Dice Game*

With the help of a dice game, we tried to simulate this working environment and used the following logic for the dice game:

(a) We have a sufficient number of patients before the registration desk for treatment.
(b) Nursing care unit has two options (psychosocial and psychiatric evaluation). Pick slip from bowl for first dice and then again dice for second unit.
(c) Similarly, the psychosocial evaluation unit have two options, again pick slip for the psychiatric and therapy unit. The first dice will be for the first slip and the second for the other department.
(d) From all the departments, we can only move six patients per hour to the next department.
(e) There is no limit for patients in the queue before any workstation.
(f) Play this game for five days
(g) Registration desk will be open for six hours only.

Figure 2a shows the real setup of the healthcare unit for the dice game and Fig. 2b represents the flow of patients during the dice game. We simulate for five days and record values on the Excel file for further analysis. Here, Fig. 3 presents the values for day 1 to day 5 for every hour. In which inventory, dice number, and number of patients treated data are available for conceptualization of the concept of a dice game.

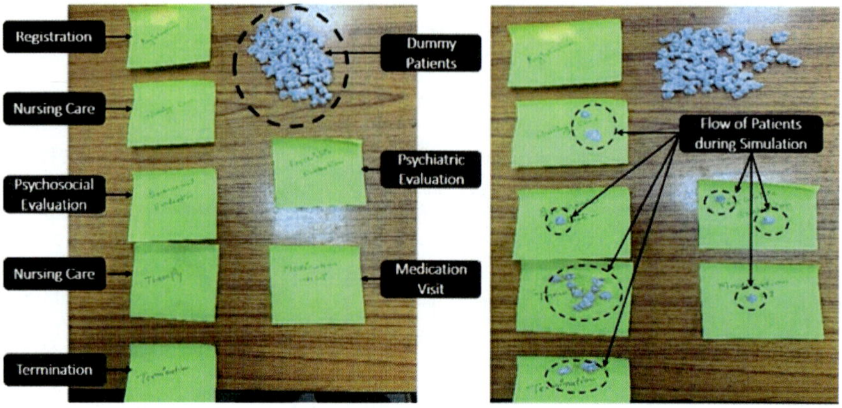

Fig. 2 **a** Setup for dice game, **b** Flow of patients during simulation

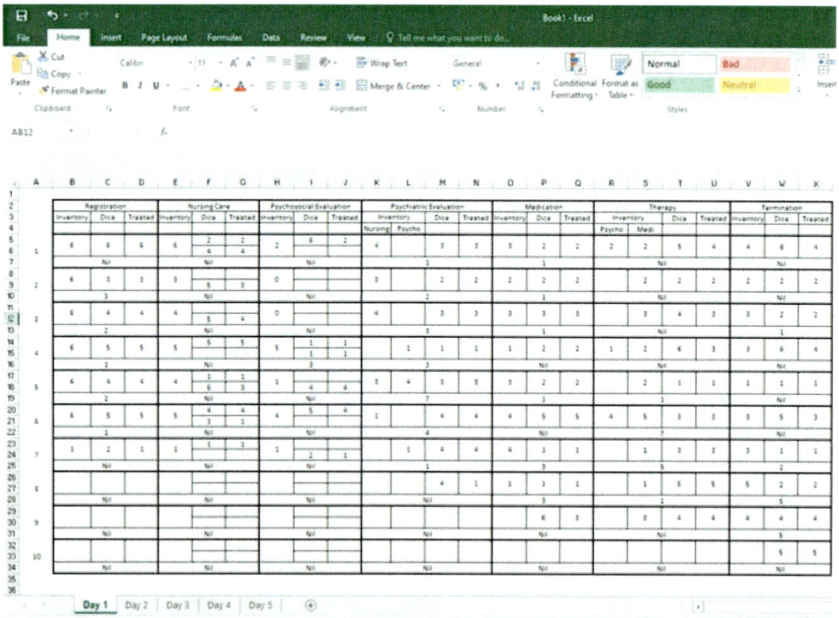

Fig. 3 Excel file of dice game

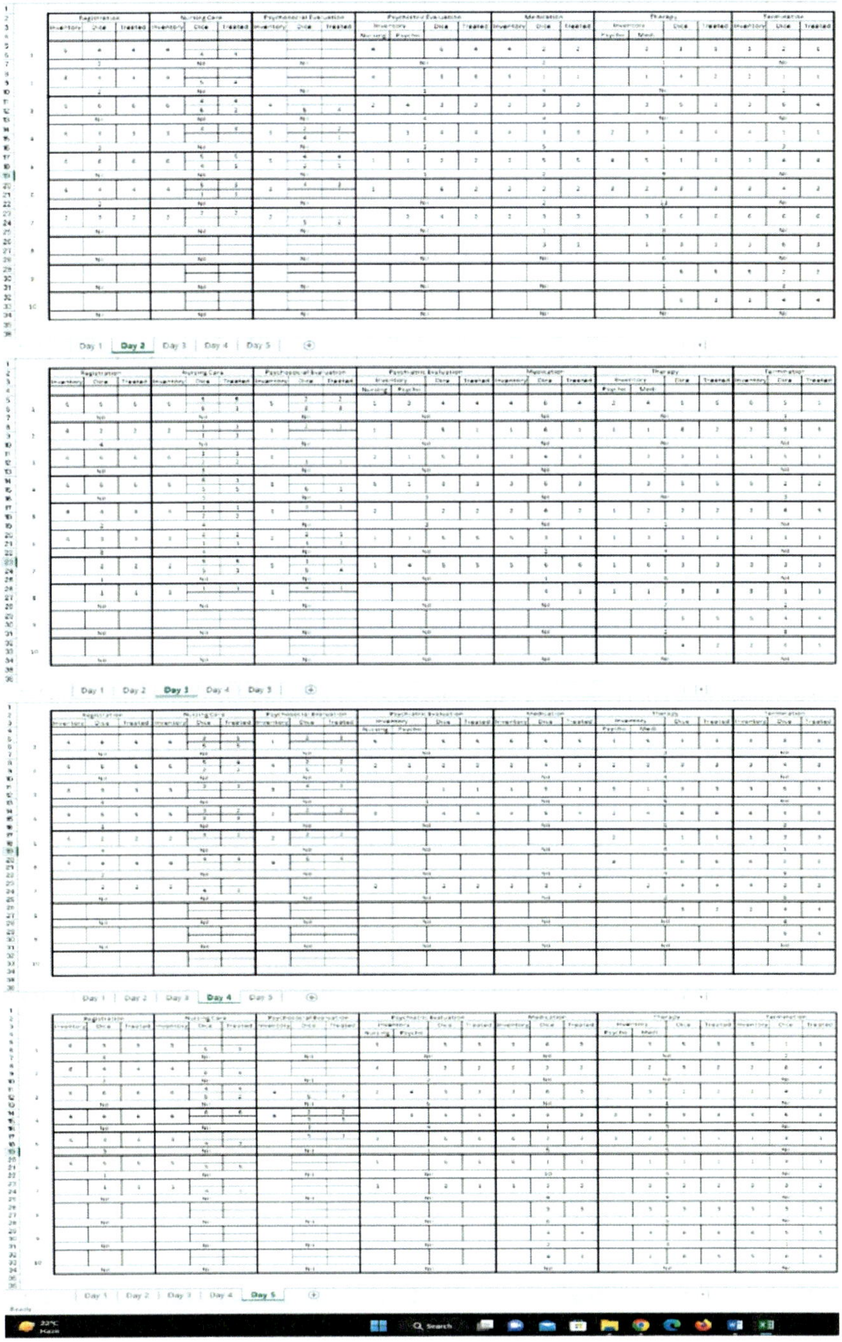

Fig. 3 (continued)

3 Results and Discussions

For identification of the constraint in this healthcare unit, we performed statistical analysis for visualization of actual bottlenecks in the system. For this monitored the inventory available before each and every department and the number of patients treated by each department.

3.1 *Inventory Analysis*

Figure 4 shows the hour-wise graphical representation of inventory for five days before every department. Initially, in the morning, there was sufficient inventory in front for registration, and the nursing care unit but from the 7th hour inventory started reducing but therapy and termination departments were with work inventory which caused delays in the departure of patients.

3.2 *Patient Diagnosed Analysis*

Mostly all the departments were not performing as required by the healthcare unit. Then, we analyzed per hourly production or performance of all the departments and found registration and nursing departments were performing well followed by therapy, termination, psychiatric, medication, and psychosocial. Figure 6 represents the average number of patients treated per day and we found for registration it was 28.6, nursing (28.6), psychosocial evaluation (14.6), psychiatric evaluation (20.6), medication (20.6), therapy (28.6), and termination (28.6) (see Fig. 5).

As per the schedule, the registration desk opens for only six hours but due to policy constraints in the system all the departments worked more than six hours which causes overtime for the employees working in those departments. Figure 7 represents the graphical representation of the working hours per day for each and every department of the healthcare unit. The termination department mostly worked 9.8 h per day followed by therapy, medication, registration, nursing care, psychiatric, and psychosocial department.

3.3 *Overtime*

Mostly, all the departments performed overtime during five days simulation and reported overburden on the employees. Termination department employees did overtime 63.34% followed by therapy (56.76%), medication (40%), registration (20%), psychiatric (20%), nursing care (20%), and psychosocial (13.33%) (see Fig. 8).

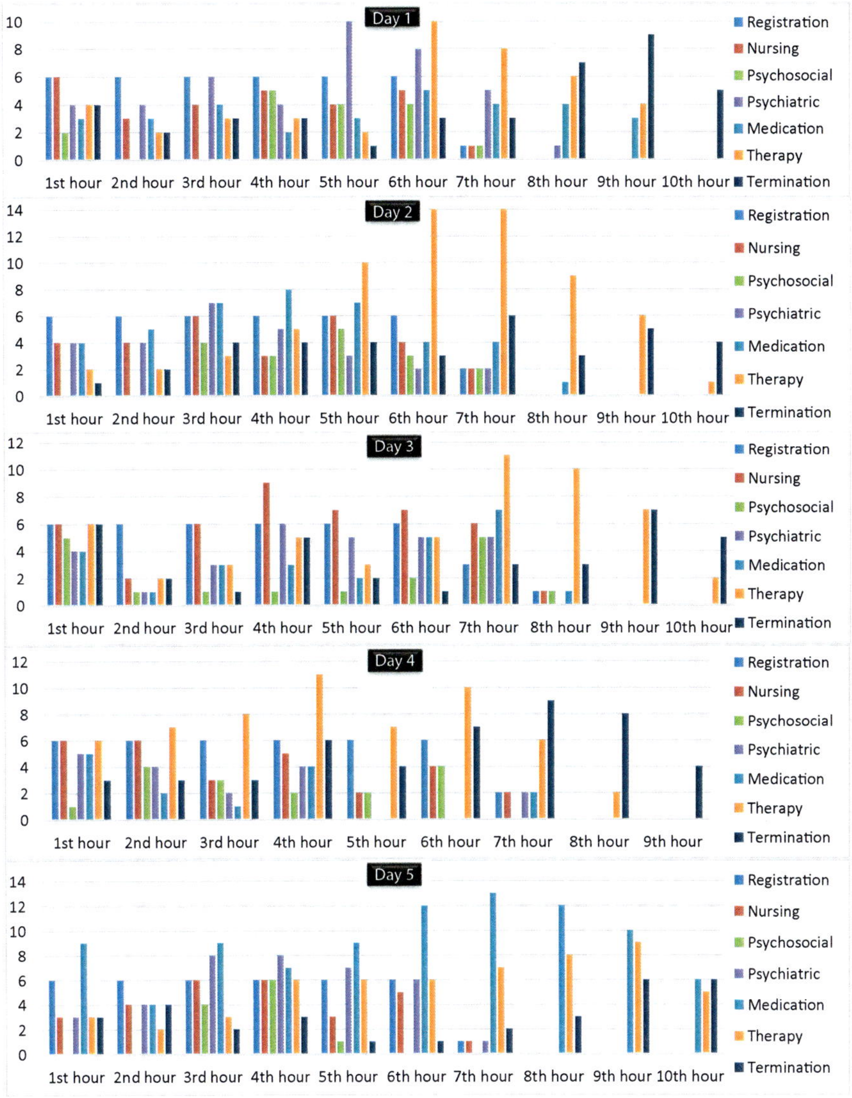

Fig. 4 Graphical representation of inventory for five days

3.4 *Resource Utilization*

Figure 9 represents resource utilization for the first six hours and reported average resource utilization for five days. For productive operations, all the resources should be properly utilized but in our case registration was maximum utilized by 74.44%

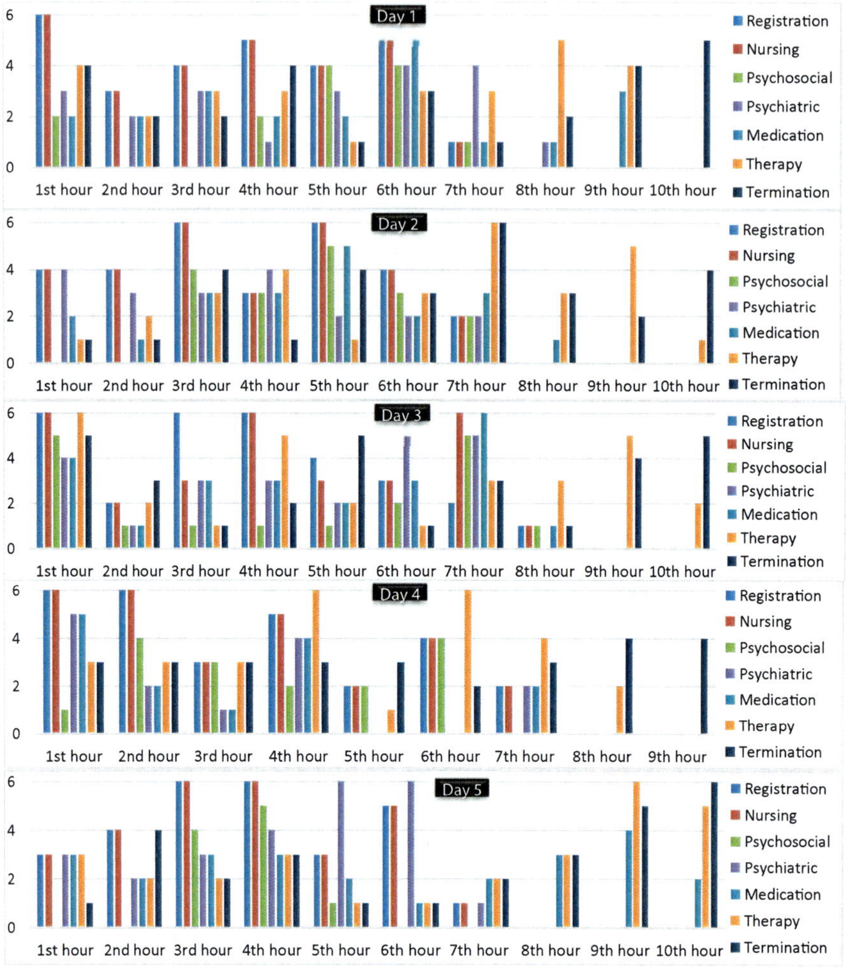

Fig. 5 Graphical representation of patient treated for five days

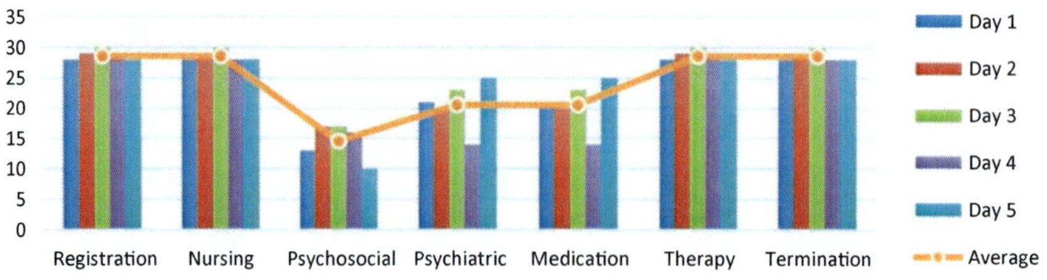

Fig. 6 Average number of patients treated per day

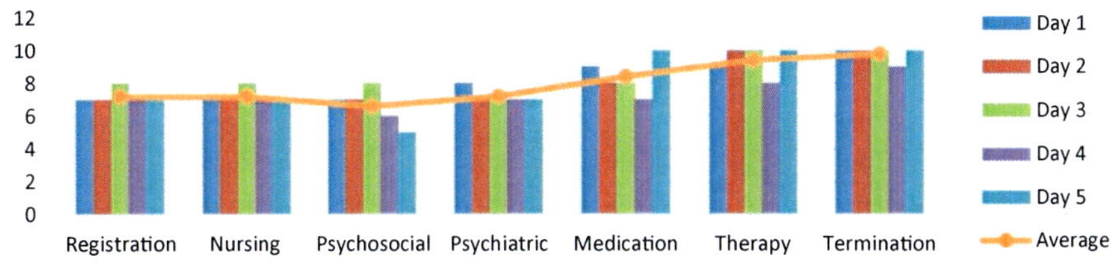

Fig. 7 Graphical representation of working hours per day

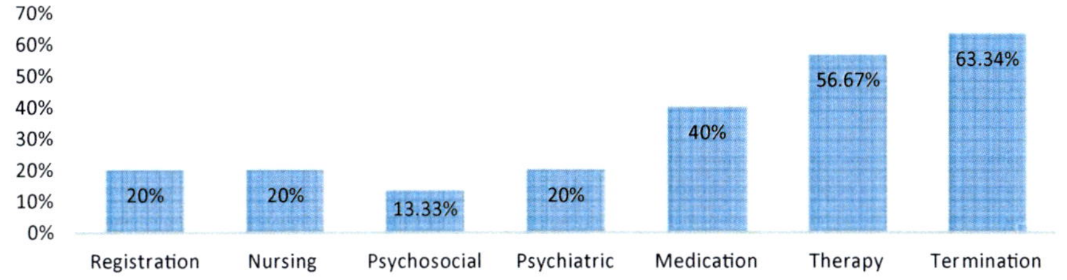

Fig. 8 Graphical representation of overtime

followed by nursing care (72.22%). Rest of the departments were utilized below 50% and due to this waiting time for patients increased.

In our simulation-based case study, we found policy as a constraint. Because the lack of triage policy was missing in the whole system resources were not utilized properly. Hence, waiting time for patients increased and reported dissatisfaction in the patients. Over time, overburdening increased dissatisfaction among employees. Hence, the performance of the healthcare unit was reduced and revenue loss occurred. According to TOC, throughput should increase, and inventory and operating expenses should decrease. But in our healthcare unit, operating expenses and inventory increased and throughput in healthcare was reduced. A proper triage policy will monitor the system and suggestions for the psychosocial department are required.

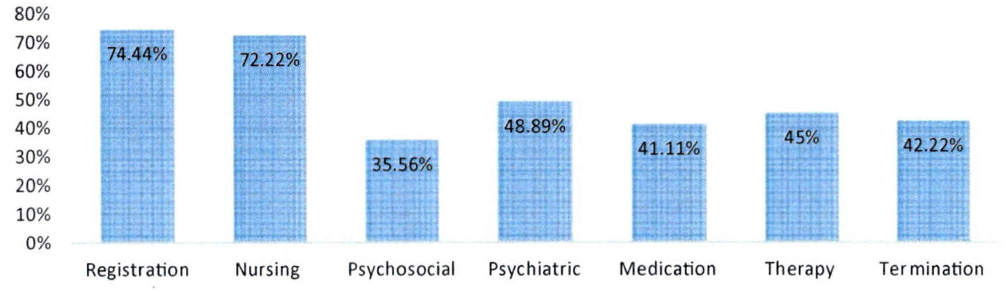

Fig. 9 Graphical representation of resource utilization

4 Conclusions

In this paper, the TOC mindset is used for the visualization of the complex healthcare system and identified constraints of the system using a dice game. TOC philosophy proposed measurement techniques totally different from traditional performance measurement techniques. A dice game was performed to generate real-life situations in complex environments and many service quality parameters scored negative scores in this system. Patients and employees were dissatisfied in this healthcare unit. This game simulates for only five days in this paper but when more simulations are performed in the future more precise results can be obtained. In the future, academicians and practitioners can find the net profit, return on investment, inventory turnover, and productivity of the healthcare unit.

References

1. Goldratt EM, Cox J (2004) The goal: a process of ongoing improvement (3., rev. ed., 20. anniversary ed). North River Press
2. Goldratt EM (1994) It's not luck. North River Press
3. Goldratt EM (1997) Critical chain. North River Press
4. Dettmer HW, Dettmer HW (2007) The logical thinking process. A systems approach to complex problem solving. American Society for Quality
5. Gupta M, Andersen S (2018) Throughput/inventory dollar-days: TOC-based measures for supply chain collaboration. Int J Prod Res 56(13):4659–4675
6. Banerjee A, Mukhopadhyay SK (2016) A contemporary TOC innovative thinking process in the backdrop of leagile supply chain. J Enterp Inf Manag
7. Moreira MRA, Castaño JD, Sousa PS, Meneses RFC (2014) Applying Goldratt's framework to the banking system. Period Polytech Soc Manag Sci 22(2):107–117
8. Golmohammadi D, Mansouri SA (2015) Complexity and workload considerations in product mix decisions under the theory of constraints. Nav Res Logist 62(5):357–369
9. Bhattacharya A, Vasant P, Sarkar B, Mukherjee SK (2008) A fully fuzzified, intelligent theory-of-constraints product-mix decision. Int J Prod Res 46(3):789–815
10. Chou YC, Lu CH, Tang YY (2012) Identifying inventory problems in the aerospace industry using the theory of constraints. Int J Prod Res 50(16):4686–4698
11. Trojanowska J, Dostatni E (2017) Application of the theory of constraints for project management. Manag Prod Eng Rev 8
12. Dalci I, Kosan L (2012) Theory of constraints thinking-process tools facilitate goal achievement for hotel management: a case study of improving customer satisfaction. J Hosp Market Manag 21(5):541–568
13. Reid RA, Cormier JR (2003) Applying the TOC TP: a case study in the service sector. Manag Serv Qual: Int J
14. Gundogar E, Sari M, Kokcam AH (2016) Dynamic bottleneck elimination in mattress manufacturing line using theory of constraints. Springerplus 5(1):1–15
15. Umble M, Umble E, Murakami S (2006) Implementing theory of constraints in a traditional Japanese manufacturing environment: the case of Hitachi tool engineering. Int J Prod Res 44(10):1863–1880
16. Taylor LJ, Nayak S (2012) Goldratt's theory applied to the problems associated with an emergency department at a hospital. Adm Sci 2(4):235–324
17. Cox III JF (2021) Using the theory of constraints to create a paradigm shift in organization performance at a large primary care provider practice. Health Syst 1–34

18. Ammar S, Wright R (1999) Experiential learning activities in operations management. Int Trans Oper Res 6(2):183–197
19. Umble E, Umble M (2005) The production dice game: an active learning classroom exercise and spreadsheet simulation. Oper Manag Educ Rev 1:105–122
20. Jackson P (1996) The cups game, NSF product realization consortium module description. Cornell University, Ithaca, NY

Development of Paddy Transplanter Machine Using Low-Cost Materials

Subhash Waghmare, Sagar Shelare, Nischal Mungle, Vinod Sakhare, and Mahendra Dhande

Abstract Agriculture is the foundation of India and can play an important role in the Indian economy. Rice is a commonly consumed staple food, particularly in Asia, which accounts for about 90% of total rice production all over the world. Rice is a key resource in India and essential for all human beings, but farmers are facing more paddy cultivation problems. The two traditional methods of establishing rice are direct seeding, rice, and transplanting. Rice is cultivated manually in direct seeding while rice is cultivated by the computer in transplantation. Farmers face numerous problems in the manual cultivation of paddy, such as backbone pain and wages for workers. Due to greater yield and less weed growth, the transplanting method is more common among farmers compared to direct-seeded rice. A compact, low-cost paddy transplanter was developed and tested to eliminate the above-mentioned drudgery. Field trials were performed in the Bhandara district of Maharashtra. In terms of field efficiency, field capacity, and fuel consumption, the performance of the produced paddy transplanter was found to be very satisfactory compared to the conventional manual transplanting process. After the reduction of the cost of the machine, this study indicates a simple design and the holder will make the job more efficient. In comparison to the manual process of paddy transplantation, the system was found to be farmer-friendly and feasible in terms of time, money, and labor requirements.

Keywords Rice · Paddy transplantation · Manual seeding · Agriculture · Plantation

S. Waghmare · S. Shelare (✉) · V. Sakhare · M. Dhande
Department of Mechanical Engineering, Priyadarshini College of Engineering, Nagpur 440019, India
e-mail: sagmech24@gmail.com

N. Mungle
Department of Mechanical Engineering, Yeshwantrao Chavhan College of Engineering, Nagpur 440019, India

A. Sachdeva et al. (eds.), *Recent Advances in Operations Management and Optimization*, Lecture Notes in Mechanical Engineering,
https://doi.org/10.1007/978-981-99-7445-0_6

1 Introduction

Indian rice is the world's largest cereal [1]. It uses up roughly 23.3% of the country's arable land [2]. In terms of both rice area covered (44.6 million hectares) and rice output (approximately 142.5 million tones), India is next to China [3, 4]. The second most important crop in Maharashtra is rice, which is cultivated over a region of 14.99 lakh tones and yields 32.37 lakh tones of raw rice per year [5, 6]. The Konkan area saw the highest yields, with 2.56 tones per ha (3.65 tones for raw rice) [7, 8]. Either direct sowing or transplanting can be used to successfully grow rice [9]. Planting seeds on soil that has been saturated by water is a widespread practise throughout Asia [10]. Growing old paddy takes a lot of time and energy and costs a lot of money. Half of all production costs are attributable to the processes of puddling and transplanting [11]. Transplanting requires between 40 and 50 man-days per ha. Planting seeds manually results in poor seed placement and spacing efficiency, as well as considerable back pain for the farmer, limiting the total area that may be sown [12–14]. Most of our farmers cannot afford the high price of imported planters [15, 16]. If peasant farmers' labour can be made easier, especially during the planting process, significantly more food, especially grains, can be produced [17–19]. A seed planter's effectiveness may be maximised by carefully tailoring its design and component choices to the specifics of a certain crop [20, 21]. There has been a recent uptick in demand for rice trans-planter machines due to their ability to sow seeds in a systematic, efficient fashion [22, 23]. The time and energy saved by using a rice trans-planter to sow seeds is immeasurable [24–26]. Farmers, who tend to be economically disadvantaged, are the main demographic that relies on these kinds of tools [27, 28]. Hence a less expensive machine is required for rice transplanting for the farmers [29, 30]. Although various attempts are made for providing various rice transplanting machine.

When designing and assembling a seed machine, it is important to keep the plant's demands in mind and work within the aforementioned constraints [31, 32]. The demand for rice cultivation is growing today because it has different characteristics of sowing in a good and orderly manner [33–35]. Sowing with rice trans-planter saves too much human effort [36]. The people who use this type of equipment are farmers and have a poor economic background. Therefore, a less expensive machine is needed to replant rice for farmers. Although various efforts have been made to supply various rice implants [37].

2 Materials and Method

2.1 *Overview of Research*

The primary goal of this innovation is to create a lightweight and portable rice transplanting machine. If you want to seed paddy more precisely and in less time, you need a machine like the one provided by this technology. Additionally, the current idea seeks to lessen the workload of the farmer and cut down on the machine's total price. Kumbhar et al. [3] In this paper, the author explains the detailed history of paddy transplantation from manually to machined and fabricated the paddy translator machine, this tool is invaluable to the modern farmer, increasing his efficiency and output. The machine was built with the resources at hand. The machine boasts better efficiency and cost savings than comparable commercial equipment. An analysis of the machine's efficiency shows that it cuts down on both translation time and manual labour. It was discovered by Rajamanickam et al. [38]. Pradhan et al. [39] study traditional transplanting with 16 births over multiple age ranges. They investigated two-, three-row, and four-row transplanter techniques using a variety of parameters. Based on the findings of the research, it is clear that a two-row rice plantation is the optimal setup for rice production. Manually rice transplanting experiments were undertaken by G. Singh et al. at the International Rice Research Institute farm [40]. The results indicate that the transplanter is able to finish the operation at a rate of 3.4 ha/h in varying water depths over this particular area. In order to determine how much pulling force a manual rice transplanter needs, researchers Yadav and Pund [41] The height at which the force is delivered is a critical process parameter. A mat-style nursery-raising apparatus for rice planters was designed by Sharma and Singh [42]. They found that compared to manually tending a nursery, they could save their time and energy expenditures by 72.02 and 33.33%, respectively. Mat-type nursery cost for 1 ha using the produced device was Rs. 299.50, whereas the cost using the manual approach was Rs. 1608.75. According to research by Yadav et al. [43] paddy transplanters save labour costs by roughly 85% compared to hand transplanting. Kumbhar et al. [3] details of the manual rice translator were laid out in his study. A shortage of experienced agricultural labourers persists during peak transplanting seasons, despite the widespread notion that surplus agricultural labour is available in India. The yield decreases if this process is delayed. This highlights the critical necessity to automate the process. In puddled wetland conditions, using peregrinated paddy seeders produces the same yield as hand transplanting but with a 75–80% decrease in labour. Manually and self-propelled transplanters cut down on expenses by around half. The group led by Harisch and Srivastav and including Using a 3:1 bar linkage, the authors of this work were able to increase the chain drive speed from the driver to the driven shaft, causing the machinery to vibrate at a predetermined angle throughout the rice transplanting operation. The rice plant is picked up easily and solely in a downward motion thanks to the sowing finger's ergonomically designed. This project, titled SangappaHadapad, aims to create a hand-operated rice transplanter for use by small-scale Indian paddy farmers. There is an immediate need to create farm machinery for extremely labour-demanding agricultural tasks in India, where more than 80% of rice landholdings are smaller than two hectares, and where labour availability is quickly reducing. It has a pedal-operated mechanical gear that turns the linkages, a tray to contain the seedlings, and the linkages themselves. It may uniformly seed two rows of soil at once. The advantages of mechanical transplanting of rice are discussed, as well as some of the operational and administrative concerns that must be handled, in a study by Amit Mishra et al. This user-friendly, step-by-step guide will enable extension workers, service providers, and forward-thinking farmers utilise mechanical transplanting with or without puddling. A self-propelled rice transplanter is used in the process of mechanically transplanting immature rice seedlings that have been cultivated in a mat nursery. In the traditional method of hand transplanting, it takes between eight and twelve workers to cover an acre. However, with the help of a self-propelled rice transplanter, a team of three workers may plant as much as four acres of rice in a single day. As part of a team led by Akshay Ichal, he designed a novel rice planter machine for establishing rice seedlings in paddy fields [44]. A mat-style rice nursery with a seed-starting tray that resembles a shed roof. There is a seedling tray shifter that moves the seedling tray like a carriage of typewriters, and there are several pickup prongs that pick up a seedling from a mat-style nursery on the seedling tray and plant it in the ground as if it were held between human fingers. When compared to hand transplantation, machine transplanting using rice transplanters saves a tone of time and effort. It almost doubles the daily planting potential from 700 to 10,000 m^2.

2.2 *History of Rice Transplanter*

2.2.1 Manual Transplanters

A fork was forged onto one end of a rod made of iron. An ergonomic wooden grip was attached to the rod. About 45 cm might be subtracted from the tool's total length. The fork could pick up anywhere from two to four seedlings at a time, and then push them into the muddy dirt. Compared to hand transplanting, the planting rate might be raised by around 20% with expertise. Squatting was also diminished due to the device's use. The operator of the transplanting device had no tactile feedback as to the depth of the planting, which was one of its drawbacks. Small plates at right angles to the rod helped the operator see how deep they were planting. It wasn't until 1955 that a gravity-type hand-operating transplanter was designed and tested for use in the paddy transplantation process [8, 9].

2.2.2 Animal Drawn Transplanter

The machine consisted primarily of a float onto which the rest of the mechanism was bolted, a long sloppy tray, a wire comb attached lengthwise on the tray, a rotor shaft on which six discs were clamped in various ways, and a ground wheel. When activated, the mechanism caused the tray to rock back and forth. The seedling tray was filled with seedlings, and the spinning clamps dug holes in the soggy dirt and placed the seedlings there. The device that releases plants into the soil during planting was given a nudge in the upward direction at the appropriate time [12]. According to data collected from trials conducted in the field, the draught was too much for a pair of ordinary animals, the seedlings planted were lying flat on the earth, and the planting forks frequently became clogged, leading to inconsistencies in the planting process [13].

2.2.3 Self-propelled Transplanter

For transplanting, this equipment required seedlings to be between 25 and 30 days old. Soil needed at least 48 h to settle. In a field with a capacity of 0.08 ha/h and a forward speed of 1–1.2 km/h, the distance between plants varied from 8 to 14 cm and the row spacing was 20 cm [16]. According to the study, labour costs may be drastically cut by using a self-propelled rice transplanter to sow seeds (6–8 rows at a time). These transplanters are currently best suited to larger farms on flat terrain and are not yet appropriate for use in mountainous areas. They need to be germinated on a 15–22-day mat on a plastic-covered raised bed [17]. There was no way for a transplanter to function properly if it had to be manually operated or pulled by bullocks. To overcome the constraints of the Konkan area, a smaller, self-propelled transplanter (two or four rows) is needed for use with mat-type nursery seedlings on small farms.

2.2.4 Preparation Process for Paddy

Land preparation process: Tilling the soil with a tractor to loosen it up is one method of land preparation. After that, water is added to the field, and the area is ploughed again with a tractor equipped with a cage well. To achieve a flat surface, a leveler is employed [15].

Seedling preparation process: When preparing seeds, they are often sown all around a nursery. It is common practice to transfer seedlings after they have reached maturity. After 24 days at the nursery, the young seedlings are transplanted to their final locations, where they are given more room to grow.

Traditional/hand transplanting or manual transplanting: Planting rice requires lengthy hours of standing in a muddy field and bending over. Mechanical approach: Using the self-propelled rice transplanter, farmers are starting to use transplanting.

2.3 Development of Paddy Transplanter

Mechanisms essential to Rice transplanting machines include the paddy seedling, planting unit, power transmission, and attachments. Figure 1 shows the basic mechanism used in Paddy Transplanter.

Depth of planting: Root development and survival in a waterlogged environment both depend on how deeply you plant. The apparatus will plant seeds at a depth of 50 mm below the surface.

Designing of tray: The seed pad and tray will be transported to the planting arm. The breadth, length, angle, and velocity of movement are the foundational parameters used to build the tray mechanism. Each planting hand has its own set of two trays since two plant rows must be planted simultaneously. The volume collected from the planting finger at a time determines the amount the tray moves per one planting of arm. The tray's travel distance is assumed to be 5 mm. The seed mat has to fall to the end of the tray so that it may be continuously fed to the planting arm. While a steeper angle makes it easier to feed the seed mat to the transplanting arm, it also causes the nursery to collapse at the tray's end, making it more difficult for the arm to remove the plants.

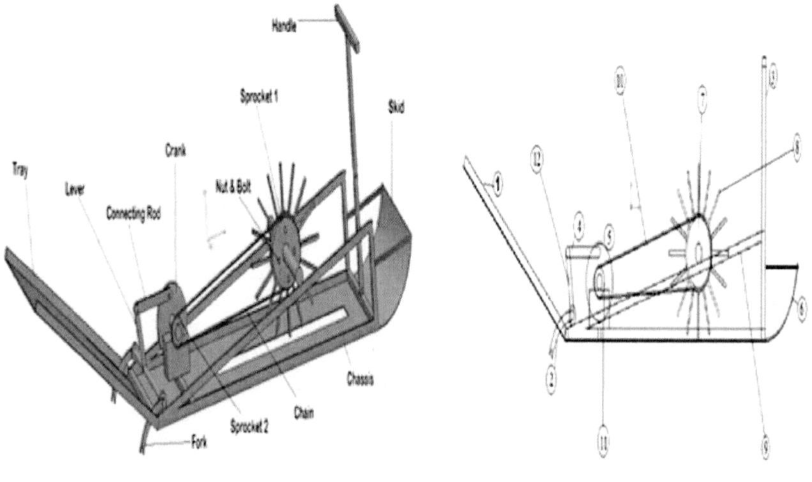

Fig. 1 Various views of the basic mechanism used in paddy transplanter

Table 1 Components dimension of paddy transplanter

Part no.	Ys (MPa)	Uts (MPa)
1	Tray	600 × 575
2	Fork	300 × 825 × 17.5
3	Handle	887.5 × 25 × 25/250 × 25 × 25
4	Connecting rod	135 × 209 × 4
5	Crank	140 mm
6	Skid	L = 200 mm
7	Sprocket 1	T = 42, PCD = 162, t = 12
8	Wheel	115 × 29 × 4, ϕ250
9	Frame	775 × 525 × 25, 775 × 205
10	Chain	L = 1050, St. 39
11	Sprocket 2	T = 15, PCD = 122, t = 10
12	Lever	75 × 29 × 4, ϕ17.5, L = 575
13	Shaft	575, ϕ17.5

Power transmission system: The arm is rotated by a sprocket and chain system powered entirely by human muscle power through a foot pedal. The results show that the reduction ratio is 2.8, which is rather significant (see Table 1).

2.4 Construction of Paddy Transplanter

The construction of the transplanter includes chassis, tray, skid, handle, sprockets, chain, crank and other components. The chassis is made up of mild steel of 1 mm thickness. The dimensions of the chassis are 775 × 525 × 25 mm^3. A fixed triangular frame is connected to the chassis of dimensions 775 × 230 mm^2 and inclined at an angle of about 250 to the horizontal. The bottommost part of the tray is connected to the chassis through welding. The tray consists of two compartments. At the bottom of the tray are two forks whose function is to sow the paddy into the field up to a depth of 50 mm. The distance between two forks is calculated to be 575 mm. The fork is connected to the shaft which is further connected to the crank which operates on four-bar mechanism. The four-bar mechanism is driven by chain and sprocket drives. The wheel drives the bigger sprocket which further drives the smaller sprocket. There are two sprockets used here. One bigger sprocket of diameter 162 mm, 42 teeth, and the smaller one of diameter 60 mm and 15 teeth. The chain consists of total of 78 rollers and a chain length of 1050 mm. The handle is fastened to the skid via nut and bolt. The proposed rice transplanting machine is cost efficient, small and portable-type transplanter machine as shown in Fig. 2. Herein the wheel drives the chain and sprocket mechanism. The chain and sprocket mechanism further drives the crank. The crank is connected to the shaft through a connecting rod. In short, the four-bar mechanism helps in sowing the paddy in the ground. The shaft is further connected

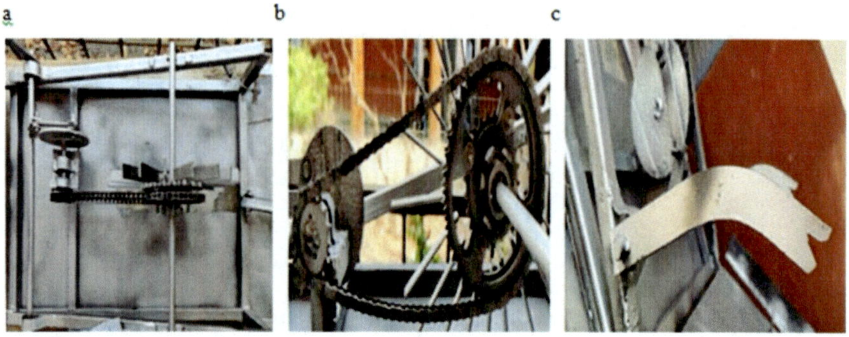

Fig. 2 **a** Main frame **b** sprocket and chain **c** fork

to the two forks which help in sowing the paddy. The fork is driven by the shaft. A chain and sprocket set up on the wheel provides the drive for the shaft. Plants for making paddy are stored in a tray and allowed to fall to the bottom by gravity. The fork which is attached to the shaft picks up the seedlings and sows them into the ground up to a depth of 50 mm. The handle attached to the skid helps in pulling the entire machine. The preferred embodiment comprises of chassis, tray, skid, handle, sprockets, chain, crank, lever, shaft, frame, fork, wheel, and connecting rod.

Main Frame: It is the part of the transplanter onto which the transplanting mechanism, chain, and sprocket are mounted. A four-bar mechanism, a rectangular frame for the tray, and supports for the transplanting mechanism make up the primary frame. The pole was fashioned from a piece of mild steel flat 4 mm thick, and it had two notches where the transplanter and its power supply could be attached. The tray was held up by a rectangular frame fashioned out of an M.S. angle that measured 25 cm by 25 cm by 3 cm. The tray can only be moved in a horizontal plane thanks to the wheels attached to the rectangular frame.

Sprocket: The transplanter consists of two sprockets. Bigger Sprocket (sprocket 1) of diameter 16.2 cm and Smaller sprocket (sprocket 2) of diameter 12.2 cm. Sprocket 1 consists of 42 teeth and sprocket 2 consists of 15 teeth. Both the sprockets in combination with the chain give a reduction in speed of 2.8:1.

Picking—Cum—Transplanting Mechanism: This mechanism's job is to pluck a certain number of seedlings from the tray. Additionally, it raises these seedlings to the soil's surface in a standing position. The seedlings are transplanted onto puddle soil using this approach. It has a pair of forks and a pair of rocker arms. A four-bar system controls the fingers. In each full turn of the wheel, the fork plants two seeds.

Crank, Lever, Connecting Rod: The crank used here is made up of mild steel and is circular in shape with a diameter of 14 cm. The crank made up of M.S. is further connected to a connecting rod of rectangular shape and of dimension $13.5 \times 2.9 \times 0.4\ \text{cm}^3$. The connecting rod is connected to the lever via a nut and bolt. The lever used here is made up of mild steel of dimension 7.5 × 2.9 × 0.4 cm^3. The lever is welded on the shaft to which the two forks are connected.

Seedling Tray: Steel sheet was used to create the seedling tray (mm). It was split in half so that two halves of rice could be stored separately. The compartment size was maintained at 60 cm^2 by 48 cm^2. Paddy seedlings move horizontally after each stroke of the transplanting arm, at which point they are transplanted by the fork. A 45-degree angle was cut into the frame and a rectangular frame was used to hold the tray from behind. The tray's distance from the mainframe was maintained at 300 mm.

Chain: The power transmission unit is the component that links the drive wheel and planting device to the chain and sprocket. The chain that was utilised has dimensions of a width of 1 cm, a pitch of 1.8 cm, and a total length of 105 cm. 78 rollers make up the chain in total.

Nut and bolt: The total number of nuts and bolts used is ten, and of those, six are made of cast iron and have the number 33 on their bolts. The other four nuts and bolts have the normal number 30, and they are also made of cast iron.

Handle: The transplanter may be moved using the single handle located on its front. The chain and sprocket system is operated by a lever located on the handle, which also serves as the operational lever. The transplanter's many parts are made and put together. The designed transplanter has a total dimensions of 91.25 × 25 cm.

3 Results and Discussions

The compact paddy transplanter machine was designed and developed. The fabricated model of the paddy transplanter machine is as shown in Fig. 3 in which Fig. 3a shows side view Fig. 3b shows the top view and Fig. 3c shows the rear side view.

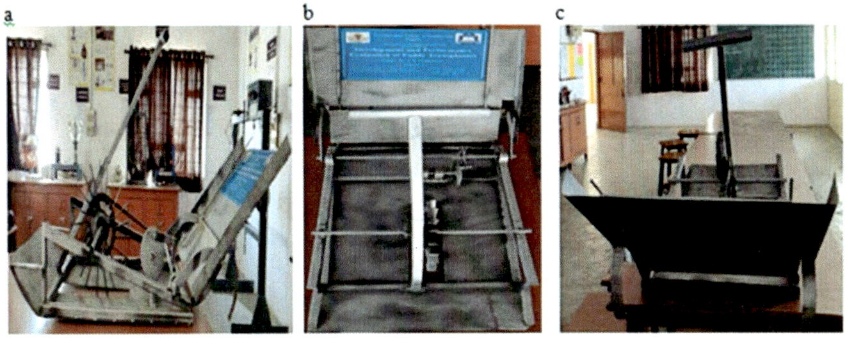

Fig. 3 Fabricated model of paddy transplanter **a** side view **b** top view **c** rear side

Table 2 details the numerous components that make up the paddy transplanter machine, including the materials they are made of, their dimensions, and the machining method that was applied. The machine that was invented is both environmentally friendly and inexpensive, making it of great service to the farming community.

Table 2 Parts list of fabricated paddy transplanter

Part name	Qty (no.)	Type of material required	Dimensions in (mm)	Machining process used
Fixed frame/ chassis	01	Hollow rectangular tubes	775 × 525, Section 25 × 25, t = 3	Welding
Triangular frame (attached to chassis)	01	Hollow rectangular tubes	Triangular shape 775 × 230, t = 3, Θ = 25°	Welding
Tray	01	Sheet	600 × 575, t = 1	Welding, Milling
Wheel plate	21	Rectangular	Stips 115 × 29 × 4, No. 21, ϕ250	Welding
Sprocket 01 (bigger)	01	Flat circular	Teeth = 42, PCD = 162 Thickness = 12	Fixed
Sprocket 02 (smaller)	01	Flat circular	Teeth = 15, PCD = 60 Thickness = 10	Fixed
Crank (circular disc)	01	Flat	ϕ140, t = 10	Drilling
Connecting rod	01	Flat	135 × 25 × 4	Drilling
Lever	01	Flat	(a) 85 × 25 × 4 (b) ϕ17.5, L = 575	Welding and Drilling
Chain	01	Single-stranded	No. of strands = 39 Pitch = 18.75, L = 1050	Fixed
Nut, bolt, visor	04	Iron	Number 33	Fixed
Bearings	05	Ball bearing	Inner diameter = 17.5	Fixed
Rod	02	Cylindrical	L = 575, ϕ17.5	Welding
Skid	01	Curved	525 × 200 × 4, R300	Welding, Milling
Shaft	02	Cylindrical	L = 575, Φ17.5	Welding
Handle	01	Flat rectangular	(a) 887.5 × 25 × 25 (b) 250 × 25 × 25	Connected

The performance evaluation of the manual paddy transplanter was carried out at Taluka Nagbhid district, Chandrapur. The experimental soil of the site was clay mud soil. An experiment field was laid out in a completely randomised design (CRD) in an area of 1196 m^2 with a plot size 53 × 7.5 m (L × W) of paddy fields located in Chandrapur district. In a completely randomised design, the treatments were taken as paddy transplantation operation and manual selection as shown in Table 3. Also, Tables 4 and 5 show the operational parameter and criteria for the selection of operational parameter, respectively.

The field was arranged using a common tillage exercise, which is initial ploughing two times with a cultivator, followed by pudding and leveling was done once with a rotavator in 100 mm depth of water with a tractor. During the field performance machine parameters, operational parameters and crop parameters were calculated. Soil-bearing seedlings ready by the traditional practice of nursery rising were used for the evaluation. The seedlings rice of 21 days were used for the transplanting

Table 3 Experimentation dimensions of the paddy transplanter

Parameter of transplater	Specification
Power source	Manually
No. of rows of paddy	2
Row to row spacing, mm	240
Maximum plant-to-plant spacing, mm	330
Type of planting mechanism	Crank type
Overall weight of paddy transplanter	20 kg
Planting depth, mm	180 mm
Type of wheel	Metal wheels with a thick rim
Seeding type	Try type
Planting length unit, mm	1150
Planting width unit, mm	1950
Planting Height unit, mm	960

Table 4 Paddy machine transplant and operational parameter

Sr.no.	Parameters	PTO and other parameter selection
1	Plot area, m^3	36.38
2	Total time taken, h	0.08
3	Speed of operation, kmph	0.23
4	Effective filed capacity, ha h^{-1}	0.023
5	Theoretical filed capacity, ha h^{-1}	0.045
6	Field efficiency, %	13.68
7	Wheel slip, %	1.67
8	Filed machine index, %	13.68

Table 5 Parameter selection during paddy transplanting field operation

Sr. no.	Parameter	PTO and parameter selection
1	Hill spacing, mm	85.50
2	Row spacing, mm	208.60
3	No. of seedlings per hill	1
4	Depth of transplanting, mm	42.3
5	Missing hill, %	5.2
6	Floating hills, %	3.1
7	Buried hills, %	0.43
8	Damaged hills,%	0.02
9	Standing angle, for seedlings planted	82

operation and the height of the seedling at the time of transplanting was 18 cm. The standing water at the time of transplanting was maintained at less than 20 mm [24]. The manual paddy transplanting with 2-row paddy transplanter was made in a completely randomised design (CRD) to find optimum plant pudding than manual labour. Based on the study, the following results were drawn that two- and three three-row manual transplanters was found to be improved in performance on the source of total unproductive hills (6.89%), planting efficiency (93%), and field efficiency (72%).

4 Conclusions

The major conclusions drawn from the study are the following:

The paddy transplanter machine provides a small and portable type paddy transplanter machine. It also facilitates sowing paddy with more accuracy and with less time consumption.

The proposed transplanter machine comprises of chassis, tray, skid, handle, sprockets, chain, crank, lever, shaft, frame, fork, wheel, and connecting rod.

In order to improve the efficacy and efficiency of paddy cultivation, there has been a rise in the need for automation in the paddy transplanter field.

The major problem of the current transplanter is the high price and the intricate design of the mechanism. The regular repair shop just wasn't cutting it for this type of care and upkeep. In order to help farmers who are unable to work due to illness, this inspired the creation of a rice transplanter.

When a farmer's plot of land is somewhat small, a mechanised paddy transplanter can help by reducing the amount of labour required to complete the paddy transplant.

References

1. Phoksawat K, Mahmuddin M (2017) Hybrid ontology-based knowledge with multi-objective optimization model framework for decision support system in intercropping. Adv Sci Technol Eng Syst J 2:1363–1371
2. Shelare S, Kumar R, Khope P (2021) Assessment of physical, frictional and aerodynamic properties of Charoli (Buchanania Lanzan Spreng) nut as potentials for development of processing machines. Carpathian J Food Sci Technol 13:174–191. https://doi.org/10.34302/crpjfst/2021.13.2.16
3. Kumbhar S, Khot S, Mohite S, Pandit SV (2017) Design analysis and fabrication of manual rice transplanter. Int J Innov Res Sci 6
4. Rickman JF, Mussgnug F, Khanda CM, Satpathy SD, Parida N, Singla K, Kumar V, Banik NC, Iftikar W, Mishra A (2015) Operational manual for mechanical transplanting of rice
5. Xin L, Lv Z, Wang W, Zhou M, Zhao Y (2017) Optimal design and development of a double-crank potted rice seedling transplanting mechanism. Trans ASABE 60:31–40
6. Waghmare SN, Shelare SD, Tembhurkar CK, Jawalekar SB (2021) Development of a model for the number of bends during stirrup making process. In: Lecture notes in mechanical engineering. Springer, pp 69–78. https://doi.org/10.1007/978-981-15-5151-2_7
7. Kumar U, Thomas, EV (2015) Determination of force acting on the rice transplanter finger. Agric Eng Int CIGR J 17
8. Mowade S, Waghmare S, Shelare S, Tembhurkar C (2020) Mathematical model for convective heat transfer coefficient during solar drying process of green herbs. In: Advances in intelligent systems and computing, pp 867–877. https://doi.org/10.1007/978-981-32-9515-5_81
9. Shelare SD, Aglawe KR, Matey MS, Shelke KS, Sakhale CN (2023) Preparation, applications, challenges and future prospects of nanofluid materials with a solar systems in the last decade. Mater. Today Proc. https://doi.org/10.1016/j.matpr.2023.06.160
10. Belkhode PN, Ganvir VN, Shelare SD, Shende A, Maheshwary P (2022) Experimental investigation on treated transformer oil (TTO) and its diesel blends in the diesel engine. Energy Harvest Syst 9:75–81. https://doi.org/10.1515/ehs-2021-0032
11. Waghmare S, Mungle N, Tembhurkar C, Shelare S, Pathare N (2019) Design and analysis of power screw for manhole cover lifter. Int J Recent Technol Eng 8. https://doi.org/10.35940/ijrte.B2628.078219
12. Jawalekar SB, Shelare SD (2020) Development and performance analysis of low cost combined harvester for rabi crops. Agric Eng Int CIGR J 22:197–201
13. Sahu P, Shelare S, Sakhale C (2020) Smart cities waste management and disposal system by smart system: a review. Int J Sci Technol Res 9:4467–4470
14. Naidu VAM et al (2020) A review on design of components of 4 stroke engine using hybrid metal matrix. Int J Mech Prod Eng Res Dev 10:8853–8862. https://doi.org/10.24247/ijmperdjun2020842
15. Aglawe K, Yadav R, Thool S (2022) Experimental investigation of Al_2O_3 nanofluid for thermal energy management of microchannel heat sink. Transdiscipl J Eng Sci 13:203–216. https://doi.org/10.22545/2022/00218
16. Shelare SD, Aglawe KR, Waghmare SN, Belkhode PN (2021) Advances in water sample collections with a drone—a review. Mater Today Proc 47:4490–4494. https://doi.org/10.1016/j.matpr.2021.05.327
17. Glance ASAA (2014) Government of India, ministry of agriculture, department of agriculture & cooperation, directorate of economics & statistics
18. Waghmare SN, Shelare SD, Tembhurkar CK, Jawalekar SB (2020) Pyrolysis system for environment-friendly conversion of plastic waste into fuel. In: Lecture notes in mechanical engineering, pp 131–138. https://doi.org/10.1007/978-981-15-4748-5_13
19. Naidu VAM, Shelare SD, Awatade SM (2020) Design of components of 4 stroke engine using hybrid metal matrix. Int J Innov Technol Explor Eng 9:238–242. https://doi.org/10.35940/ijitee.j7509.0891020

20. Shelare SD, Thakare PS, Handa CC (2012) Computer aided modelling and position analysis of crank and slotted lever mechanism. Int J Sci Technol Res 1
21. Waghmare SN, Shelare SD, Mungle NP, Aglawe KR (2023) Development of IOT-based hybrid helium drone for flight time enhancement. In: Vasudevan H, Kottur VKN, Raina AA (eds) Proceedings of international conference on intelligent manufacturing and automation. Lecture notes in mechanical engineering. Springer Nature Singapore, Singapore, pp 329–339. https://doi.org/10.1007/978-981-19-7971-2_32
22. Pateriya RN, Datta RK (2012) Design modifications of mat type rice transplanter. Int J Adv Technol Eng Res 2:2–6
23. Shelare SD, Kumar R, Khope PB (2021) Flywheel energy application in commercial and agricultural field: a typical review. In: Lecture notes in mechanical engineering. Springer, pp 177–186. https://doi.org/10.1007/978-981-16-1079-0_19
24. Waghmare S, Shelare S, Sirsat P, Pathare N, Awatade S (2020) Development of an innovative multi-operational furnace. Int J Sci Technol Res 9:885–889
25. Waghmare SN et al (2017) A case study on improvement of plant layout for effective production. Int J Mech Prod Eng Res Dev. https://doi.org/10.24247/ijmperdoct201716
26. Shelare S, Belkhode P, Nikam KC, Yelamasetti B, Gajbhiye T (2023) A payload based detail study on design and simulation of hexacopter drone. Int J Interact Des Manuf. https://doi.org/10.1007/s12008-023-01269-w
27. Gajbhiye T, Shelare S, Aglawe K (2022) Current and future challenges of nanomaterials in solar energy desalination systems in last decade. Transdiscipl J Eng Sci 13:187–201. https://doi.org/10.22545/2022/00217
28. Waghmare SN, Shelare SD, Bahl S, Mungle N, Kumar Bagha A, Chandmal Sharma R (2022) Structural design and its analysis of two station notching tool for industrial die. Mater Today Proc 68:1335–1341. https://doi.org/10.1016/j.matpr.2022.06.360
29. Waghmare SN, Sakhale CN, Tembhurkar CK, Shelare SD (2020) Assessment of average resistive torque for human-powered stirrup making process. In: Advances in intelligent systems and computing, pp 845–853. https://doi.org/10.1007/978-981-32-9515-5_79
30. Waghmare SN, Mowade KS, Shelare SD, Mowade SW (2023) Experimental analysis of improved vortex tube. In: Optimization methods for engineering problems. https://doi.org/10.1201/9781003300731-11
31. Aglawe KR, Dhande M, Matey M, Shelare S (2022) State of the art and materials based characteristics in power converters for electric vehicles. Mater Today Proc 58:726–735. https://doi.org/10.1016/j.matpr.2022.02.384
32. Shelare S, Kumar R, Gajbhiye T, Kanchan S (2023) Role of geothermal energy in sustainable water desalination—a review on current status, parameters, and challenges. Energies 16:2901. https://doi.org/10.3390/en16062901
33. Khope PB, Shelare SD (2021) Prediction of torque and cutting speed of pedal operated chopper for silage making. In: Lecture notes in mechanical engineering. Springer, pp 243–249. https://doi.org/10.1007/978-981-16-1769-0_22
34. Shelare SD, Waghmare SN, Yenurkar MM, Jodh GR (2023) Experimental analysis of sprayer for horticulture and orchards. In: Vasudevan H, Kottur VKN, Raina AA (eds) Proceedings of international conference on intelligent manufacturing and automation. Lecture notes in mechanical engineering. Springer, Springer Nature Singapore, Singapore, pp 651–660. https://doi.org/10.1007/978-981-19-7971-2_64
35. Shende V, Mehta G, Belkhode P, Shelare S, Lokhande K, Modak J, Afsar A (2021) Design and development of concrete mixer driven by human powered flywheel motor. Int J Bridg Eng 9:31–50
36. Shelare SD, Aglawe KR, Khope PB (2021) Computer aided modeling and finite element analysis of 3-D printed drone. Mater Today Proc 47:3375–3379. https://doi.org/10.1016/j.matpr.2021.07.162
37. Bautista EU, Koike M, Suministrado DC (2001) PM—power and machinery: mechanical deep placement of nitrogen in wetland rice. J Agric Eng Res 78:333–346

38. Rajamanickam AK, Uvaraja VC. Selvamuthukumaran D, Surya K (2021) Fabrication of paddy transplanter machine. In: IOP conference series: materials science and engineering. IOP Publishing, p 12064
39. Pradhan S, Elhadad N, Chapman W, Manandhar S, Savova G (2014) Semeval-2014 task 7: analysis of clinical text. In: Proceedings of the 8th international workshop on semantic evaluation (SemEval 2014), pp 54–62
40. Singh G, Sharma TR, Bockhop CW (1985) Field performance evaluation of a manual rice transplanter. J Agric Eng Res 32:259–268
41. Yadav R, Pund S (2007) Development and ergonomic evaluation of manual weeder. Agric Eng Int CIGR J
42. Sharma SC, Singh TP (2004) Pantnagar developed a mat-type nursery raising device for rice transplanters. Agric Eng Today 28:70–72
43. Yadav R, Patel M, Shukla SP, Pund S (2007) Ergonomic evaluation of manually operated six-row paddy transplanter. Int Agric Eng J 16:147–157
44. Patil DD, Phate MR (2016) Design & development of rice planter machine. Imp J Interdiscip Res 2:1362–2454

The Effect of Lean-Green-Six Sigma Practices on Organizational Performance: A Machine Learning Approach

Lokpriya Gaikwad, Chandan Chaudhari, Sandip Kanase, Yayati Shinde, Jaydeep Patil, and Vahid M. Jamadar

Abstract With the advent of Industry 4.0, more emphasis is currently being placed on the adoption of digital technologies in all areas. The current study investigates the effect of Lean-Green-Six Sigma practices on organizational performances (i.e. Operational and Financial) in relation to developing nations. A cross-sectional data set was gathered from 120 Indian manufacturing companies, and machine learning techniques were used for data analysis and result interpretation. This study gives an enhanced comprehension of ML advances in embracing Lean-Green-Six Sigma rehearses. The results show that Lean-Green-Six Sigma rehearses emphatically influence the operational and financial performance of the organizations as the adjusted R square value of 70.6 and 68.2% with P values (≤ 0.05) satisfy the hypothesis. The results likewise offer a strategy structure for supervisors, policymakers, and makers to advance Lean-Green-Six Sigma rehearses in organizations. Although a few late studies have attempted to research the force of LGSS rehearses on maintainability execution, be that as it may, not many examinations have dissected the impact of LGSS through the reception of the ML approach with regard to arising economies.

Keywords Lean-green-six sigma practices · Machine learning · Sustainability · Manufacturing firms

---

L. Gaikwad (✉) · C. Chaudhari · S. Kanase · Y. Shinde · J. Patil · V. M. Jamadar
Department of Mechanical Engineering, Navi Mumbai, India
e-mail: lokpriya2007@gmail.com

L. Gaikwad · C. Chaudhari
SIES Graduate School of Technology, Navi Mumbai, India

S. Kanase · Y. Shinde · J. Patil
Bharati Vidyapeeth College of Engineering, Navi Mumbai, India

V. M. Jamadar
Dr. Daulatrao Aher College of Engineering, Karad, MH, India

A. Sachdeva et al. (eds.), *Recent Advances in Operations Management and Optimization*, Lecture Notes in Mechanical Engineering,
https://doi.org/10.1007/978-981-99-7445-0_7

1 Introduction

The present market elements and globalization force extreme rivalry, because of associations from different ventures endeavoring to strategically upgrade their operational greatness and the executive's ways of thinking to support the upper hand and keep up market share [1]. There exist some business improvement methods, for example, Lean, Green, value engineering, Business process reengineering, Six Sigma, and so on [2–4].

Dues et al. [5] expressed that "lean serve in as an impetus for green, meaning it works with an organization's change towards green". The expression "Green" was more connected with considering the safeguarding of the indigenous habitat during the production process. As of late Green is more named with sustainability. A business system centers on hierarchical productivity by utilizing harmless ecosystem eco-materials and working cycles [6]. Arnheiter and Maleyeff [7] argued that Lean and Six Sigma are the two current and most critical strategies that associations endeavor to carry out to further develop execution and content worldwide. Su et al. [8] asserted that there has been a dramatic development in Lean and Six Sigma throughout the long term. Examples of overcoming adversity of Lean and Six Sigma execution and resulting upgrades in organizations all through the world have been displayed in scholarly diaries [5, 9, 10].

Every strategy utilizes various methodologies and standards to impact the improvement. For instance, Six Sigma centers around lessening process variety to work on quality, while Lean, then again, focuses on disposing of waste to enhance the value of the client. There are a few endeavors to coordinate LGSS; in any case, apparently, one procedure for the most part rules the other and there is no great reason or hypothetical similarity for the reconciliation of the three strategies [11, 12].

1.1 Problem Statement

Based on research survey conducted in Indian manufacturing industries to find out the best practices of Lean, Green, and Six Sigma strategies that affect to improve operational and financial performance. The aim of the study is to build a predictive model to evaluate the performance of feature variables on the target variables, i.e. operational and financial performance through a machine learning approach.

Hypothesis statements:

H1: LGSS practices affect significantly on operational performance.

H2: LGSS practices affect significantly on financial performance.

2 Data Collection and Analysis

Data was collected through the convenience sampling process where LGSS strategies are implemented in the Indian manufacturing industry. Out of the 250 managers/executives working for Indian manufacturing plants, 120 final responses were gathered via online, in-person visits, emails, etc. However, when it came to gathering information from participants, an effective rate of 48% was noted.

For a survey, a questionnaire was prepared on a five-point Likert scale having 21 measurement items of the five constructs such as LGSS practices, Operational performance, and financial performance.

2.1 *Heat Map*

A heat map is a graphic representation of information that highlights specific attributes addressed as colors. It is utilized to imagine the correlation between factors in a dataset. The heat map commonly utilizes variety coding to show the magnitude of the correlation coefficient, with darker colors addressing stronger relationships and lighter colors addressing weaker correlations (Fig. 1).

Dissemination plots were likewise plotted and generally utilized in data examination to rapidly envision the dispersion of a variable and to check for normality or skewness. The state of the distribution plot gives bits of knowledge into the basic dissemination of the information, for example, whether it is symmetric or slanted, and whether it is unimodal or multimodal.

Data scaling procedures in ML allude to techniques for changing the features of input data to work on model execution and combination.

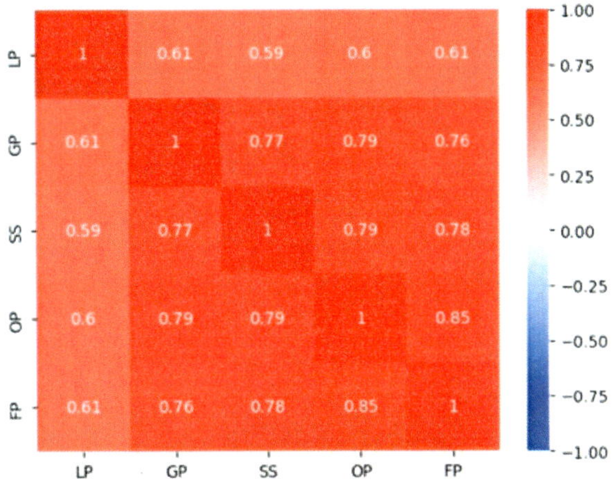

Fig. 1 Relationship of the dataset using heat map

It is vital to take note that different scaling strategies can diversely affect model execution and the decision of scaling procedure relies upon the idea of the input data and the issue being tackled. In this research paper, the MinMax Scalar method is utilized since this strategy is used since the data set isn't regularly conveyed toward mean value.

3 Results and Discussions

Model Hypothesis: Multiple Regression Performance

H1: Lean-Green-Six Sigma practices affect significantly on operational performance.

The above-stated H1 hypothesis is accepted as shown in Table 1, R Squared value for the model is 71.4% whereas the Adjusted R square value is 70.6%. The P value satisfies hypothesis H1 since this value is found to be less than 0.05. The regression equation for the below model is.

Operational performance = 0.0882 * LP + 0.316*GP + 0.3459 * SS + 0.9985.

Figure 2 indicates the implemented effect of LGSS practices on the operational performance as most of the data points are around 4.5.

H2: Lean-Green-Six Sigma practices affect significantly on financial performance.

Table 1 Ordinary least square regression result for operational performance

OLS R summary						
Dep variable:			**OP**	**R Sq:**		**0.714**
Model:			OLS	**Adj R Sq:**		**0.706**
Method:			**Least square**	F-statistic:		82.50
Date:			02 Feb 2023	Prob (F-Statitic):		7.95e-27
No.of observations:			103	Log-likelihood:		−10.638
Df residuals:			99	AOC:		29.28
Df model:			3	BIC:		39.82
Covariance type:			Nonrobust			
	Coef	Std err	t	P > [t]	[0.025	0.975]
LP	0.0882	0.054	1.630	0.106	−0.019	0.196
GP	0.3618	0.082	4.422	0.000	0.1999	0.524
SS	0.3459	0.069	4.988	0.000	0.208	0.484
Const	0.995	0.220	4.537	0.000	0.561	1.432
	Omnibus	4.737		Durbin-Watson	1.397	
	Prob (Omnibus)	0.094		Jarque–Bera (JB)	4.797	
	Skew	−0.511		Prob [JB]	0.0908	
	Kurtosis	2.728		Cond No	60.6	

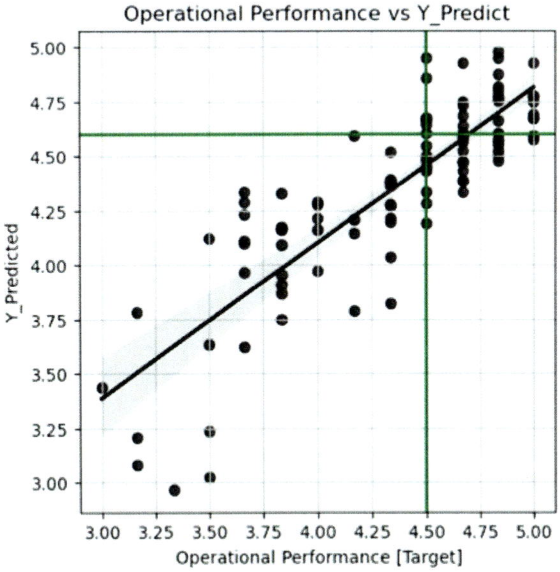

Fig. 2 LGSS practices versus operational performance

Similarly H2 hypothesis is accepted in the value shown in Table 2, R squared value for the model is 69.2% whereas the Adjusted R square value is 68.2%. The P value satisfies hypothesis H2 since this value is found to be less than 0.05. The Regression equation for the model is.

Financial performance = 0.1025*LP + 0.20772*GP + 0.3131*SS + 1.5514.

Figure 3 indicates how the data points of LGSS practices coincide with the financial performance which shows the implemented effect of practices on financial performance.

It has been observed that LGSS practices assist modern industries in enhancing sustainable performances. Most of the researcher already correlates these practices for the overall improvement of the organization in different sectors. This research is limited to Indian manufacturing industries but in the future same strategies will be implemented in different sectors and geographical areas to observe the integrated effects of these practices on firm performances.

4 Conclusions

It is observed that the Lean-Green-Six Sigma practices have positive significant effects on the operational and financial performance of the organization as the interlinking exists among these practices and defined hypotheses are proved. However, there may be variation in the contribution of the individual practices that depend upon the organizational business.

Table 2 Ordinary least square regression result for financial performance

OLS R summary						
Dep variable:			FP	R Sq:		0.692
Model:			OLS	Adj R Sq:		0.682
Method:			Least square	F-statistic:		74.09
Date:			02 Feb 2023	Prob (F-Statitic):		3.32e-25
No.of observations:			103	Log-likelihood:		−1.9855
Df residuals:			99	AOC:		11.97
Df model:			3	BIC:		22.51
Covariance type:			Nonrobust			
	Coef	Std err	t	P > [t]	[0.025	0.975]
LP	0.1025	0.050	2.060	0.042	0.04	0.201
GP	0.2772	0.075	3.685	0.000	0.128	0.426
SS	0.3131	0.064	4.910	0.000	0.187	0.440
Const	x1.5514	0.202	7.663	0.000	1.151	1.952
Omnibus		5.192		Durbin-Watson	1.661	
Prob (Omnibus)		0.075		Jarque–Bera	4.889	
Skew		−0.532		Prob [JB]	0.0867	
Kurtosis		3.079		Cond No	60.6	

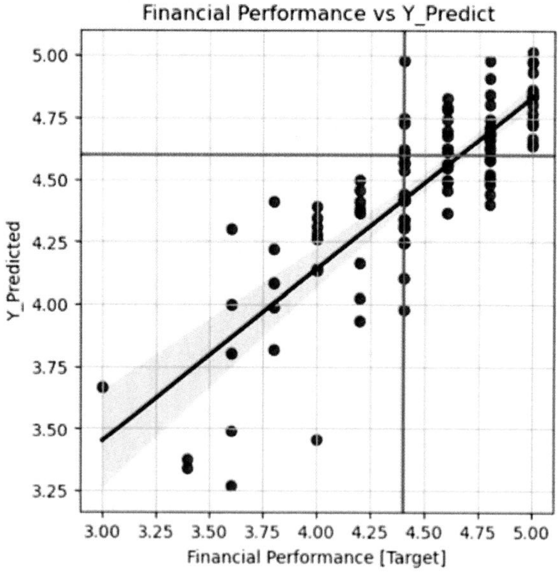

Fig. 3 LGSS practices versus financial performance

Six Sigma practices were found to be superior as compared to Green and Lean practices. Since the model efficiency is found to be 63%. Generally, for a certain organization, if the operational performance increases this will result in an increase in financial performance. i.e. more productivity and better results can be achieved by implementing a suitable model. The above machine learning model also helps us to give the best decision technique for achieving the best model fit and result.

References

1. Islam M, Karim A (2011) Manufacturing practices and performance–comparison among small medium and large industries. Int J Qual Reliab Manag 28:43–61
2. Pusavec F, Krajnik P, Kopav J (2010) Transitioning to sustainable production-part I: application on machining technologies. J Clean Prod 18:174–184
3. Andersson R, Eriksson H, Torstensson H (2006) Similarities and differences between TQM, six sigma and lean. TQM Mag 18:282–296
4. Albliwi SA, Antony J, Arshed N, Ghadge A (2017) Implementation of lean six sigma in Saudi Arabian organisations: findings from a survey. Int J Qual & Reliab Manag 34:508–529
5. Shah R, Chandrasekaran A, Linderman K (2008) In pursuit of implementation patterns: the context of lean and six sigma. Int J Prod Res 46:6679–6699
6. Dües CM, Tan KH, Lim M (2013) Green as the new lean: how to use lean practices as a catalyst to greening your supply chain. J Clean Prod 40:93–100
7. Haapala KR, Zhao F, Camelio J, Sutherland JW, Skerlos SJ, Dornfeld DA, Jawahir IS, Clarens AF, Rickli JL (2013) A review of engineering research in sustainable manufacturing research in sustainable manufacturing. J Manuf Sci Eng 135:1004–1013
8. Arnheiter, Maleyeff (2005) The integration of lean management and six sigma. TQM Mag 17:5−18
9. Su CT, Chiang TL, Chang CM (2006) Improving service quality by capitalizing on an integrated lean six sigma methodology. Int J Six Sigma Compet Advant 2:1–22
10. Pepper MP, Spedding TA (2010) The evolution of lean six sigma. Int J Qual Reliab Manag 27:138–155
11. Gaikwad LM, Sunnapwar V (2020) An integrated approach for lean, green and six sigma strategies: a systematic literature review and future directions for developing a specific framework. TQM J 32:201–222
12. Gaikwad LM, Sunnapwar V (2021) Development of an integrated framework of LGSS strategies for Indian manufacturing firms to improve business performance: an empirical study. TQM J 33:257–291

Design of an Optimized Distribution Network for the Effective Allocation of LPG Cylinders in a Closed Distribution System

R. S. Bennet Victor Samuel, N. Suriya, N. K. Aravinda Krishna, and A. Prabukarthi

Abstract Finding the best routes for numerous vehicles travelling to a collection of places is done using the Vehicle Routing Problem (VRP). In logistics, the distributor's total company performance is significantly impacted by transportation costs. Due to the scale of the Vehicle Routing Problem that needs to be solved, commercial systems employ heuristics. The Vehicle Routing Problem is employed in this study to streamline the distribution network for LPG cylinders. The distributor needs this optimization because the distribution staff's present system, which involves manually determining the delivery order, is inefficient. The goal of minimizing the overall transportation cost is achieved by formulating a Linear Programming Problem (LPP).

Keywords Vehicle Routing Problem(VRP) · Optimization · Linear Programming Problem(LPP) · Simulation · LPG distribution

1 Introduction

Common tasks in logistics management include planning, carrying out and supervising multiple procedures that ensure the effective functioning of storage and transit of goods from their point of origin to their destination. Branch and Cut method and heuristic approaches shall be used to solve vehicle routing problems [1, 2]. The two main goals of logistics management are to increase customer satisfaction and to reduce overall operating costs. Inventory models with reverse logistics and robust optimization shall be used for the LPG supply chain [3, 4]. Ant Colony Optimization (ACO) and Particle Swarm Optimization (PSO) techniques shall be used for optimizing the LPG distribution network [5, 6]. A simultaneous pickup and delivery vehicle routing problem shall be solved using the Ant Colony Optimization (ACO)

R. S. Bennet Victor Samuel (✉) · N. Suriya · N. K. Aravinda Krishna · A. Prabukarthi
Department of Mechanical Engineering, PSG College of Technology, Coimbatore, India
e-mail: bennetsamuel1999@gmail.com

A. Sachdeva et al. (eds.), *Recent Advances in Operations Management and Optimization*, Lecture Notes in Mechanical Engineering,
https://doi.org/10.1007/978-981-99-7445-0_8

algorithm [7]. Vehicle routing problems with simultaneous delivery and pickup shall be solved using heuristic algorithms [8–10].

Objective Function:

$$
\begin{aligned}
&\text{Minimize}:\ 0.8X_{01}+0.9X_{02}+0.9X_{20}+0.4X_{12}+0.73X_{13}+0.65X_{23}\\
&+0.9X_{34}+0.2X_{35}+0.8X_{45}+0.8X_{54}+0.79X_{46}+0.79X_{64}+0.66X_{56}\\
&+0.42X_{67}+0.15X_{78}+0.48X_{89}+0.31X_{810}+0.31X_{108}+0.2X_{910}\\
&+0.22X_{1011}+0.15X_{1112}+0.11X_{1213}+0.08X_{1314}+0.07X_{1415}+0.4X_{1516}\\
&+0.2X_{1617}+0.51X_{1618}+1.13X_{1418}+1.13X_{1814}+0.43X_{1718}+0.66X_{1819}\\
&+0.3X_{1920}+0.12X_{2021}+0.24X_{2122}+0.22X_{2223}+0.16X_{2324}+0.68X_{2425}\\
&+11.71X_{250}
\end{aligned}
\tag{1}
$$

The objective equation formed is shown in Eq. 1. It aims to minimize the overall distance travelled by the distribution vehicle to find out the shortest route possible. The coefficient present in each of the terms represents the distance between the two corresponding nodes represented in the subscript. Excluding the point of origin, twenty-five other delivery locations are included.

Constraints:

$$X_{01}+X_{01}>1$$

$$X_{25}>1$$

$$X_{01}=X_{12}+X_{13}$$

$$X_{02}+X_{12}=X_{23}$$

$$X_{13}+X_{23}=X_{35}+X_{34}$$

$$X_{34}+X_{54}=X_{45}+X_{46}$$

$$X_{35}+X_{45}=X_{56}$$

$$X_{46}+X_{56}=X_{67}=X_{78}$$

$$X_{78}=X_{89}+X_{810}$$

$$X_{89}=X_{910}$$

$$X_{810}+X_{910}=X_{1011}$$

$$X_{1011}=X_{1112}=X_{1213}=X_{1314}$$

$$X_{1314}+X_{1814}=X_{1415}+X_{1418}$$

$$X_{1415}=X_{1516}$$

$$X_{1516}=X_{1617}+X_{1618}$$

$$X_{1718}+X_{1618}+X_{1418}=X_{1819}$$

$X_{1819} = X_{1920} = X_{2021} = X_{2122}$

$X_{2122} = X_{2223} = X_{2324} = X_{2425}$

The first equation, which is shown by the first two restrictions, states that in order for a vehicle to depart from the origin and arrive at the destination, it must first depart. The first equation comprises two components on the left that represent potential paths from the origin to the next destination. According to the requirements, the vehicle may have departed the point if the sum of these routes is higher than or equal to 1. Similar to the previous point, the final destination may also contain a number of possible paths that, when added together, must be equal to or exceed one in order for the vehicle to have conceivably reached the location.

2 Data Collection

The customer location coordinates data are collected by traveling with the vehicle while delivering the cylinder to the customer, distance is measured from the vehicle odometer and with the help of maps, location coordinates of each customer are noted. Customer location coordinates for 25 customers and depot are presented in Table 1.

The exact distance between each customer and the depot to each customer is calculated using the client location data. The distance matrix is created to determine the order of delivery in which the vehicle travels the shortest distance. Initially, the actual total distance travelled by a vehicle along this proposed route is around 14.15 kms, and the delivery truck needs 3 hr and 13 mins to make all deliveries. An upper route planner was utilized to optimize the delivery routes taken by the delivery vehicle. It is a piece of software that is used to improve routes. The software calculated an optimum distance of 13.29 kms, saving 2 hr and 50 mins in travel time. Savings obtained in distance and time for each trip are shown in Table 2.

2.1 Cost Calculation

There are a variety of costs included in the network of supply chain. There are direct costs and indirect costs associated with any product. Fixed costs remain constant irrespective of the amount of production whereas variable costs directly vary with the quantity of production. The amount of money spent on fuel, maintenance and labor was taken into account for calculating the cost involved in the supply chain of LPG cylinders. The amount of money spent on fuel was reduced due to optimization and a significant amount of savings was obtained from the contribution of the reduction in fuel. As a reduction in operation of the vehicle, the operating costs of the vehicle also reduced in association with the reduction of labor cost. Thus a huge amount of savings was done by reducing the consumption of fuel and labor. The savings calculated from various aspects are listed in Table 3.

Table 1 Customer location coordinates

Stop	Latitude	Longitude	City	Region	Country	Zip code
Start	11.040495	76.9947465	Coimbatore	Tamil Nadu	India	641006
1	11.042382	76.9928436	Coimbatore	Tamil Nadu	India	641006
2	11.043746	76.9929156	Coimbatore	Tamil Nadu	India	641006
3	11.046819	76.9916675	Coimbatore	Tamil Nadu	India	641006
4	11.058263	76.9953715	Coimbatore	Tamil Nadu	India	641049
5	11.052679	76.9951228	Coimbatore	Tamil Nadu	India	641035
6	11.058417	77.0005676	Coimbatore	Tamil Nadu	India	641035
7	11.057209	77.0015328	Coimbatore	Tamil Nadu	India	641035
8	11.056541	77.0014106	Coimbatore	Tamil Nadu	India	641049
9	11.056412	77.0053459	Coimbatore	Tamil Nadu	India	641035
10	11.054681	77.001859	Coimbatore	Tamil Nadu	India	641004
11	11.051234	77.0041414	Coimbatore	Tamil Nadu	India	641006
12	11.049052	77.0035498	Coimbatore	Tamil Nadu	India	641006
13	11.047753	77.0023798	Coimbatore	Tamil Nadu	India	641006
14	11.048239	77.0014146	Coimbatore	Tamil Nadu	India	641004
15	11.049388	77.0001406	Coimbatore	Tamil Nadu	India	641004
16	11.047671	76.998806	Coimbatore	Tamil Nadu	India	641035
17	11.04701	76.997912	Coimbatore	Tamil Nadu	India	641006
18	11.044851	76.9986644	Coimbatore	Tamil Nadu	India	641006
19	11.047216	76.997363	Coimbatore	Tamil Nadu	India	641006
20	11.046723	76.9950707	Coimbatore	Tamil Nadu	India	641004
21	11.045196	76.9942783	Coimbatore	Tamil Nadu	India	641006
22	11.043765	76.9977713	Coimbatore	Tamil Nadu	India	641006
23	11.043626	77.0003292	Coimbatore	Tamil Nadu	India	641006
24	11.044425	77.0003067	Coimbatore	Tamil Nadu	India	641006
25	11.041846	77.0030113	Coimbatore	Tamil Nadu	India	641006
End	11.040495	76.9947465	Coimbatore	Tamil Nadu	India	641006

2.2 Simulated Results

The optimized sequence of delivery of the LPG cylinders is found using Upper Route Planner software, VROOM (Vehicle Routing Open-source Optimization Machine) simulation and FlexSim simulation software and the results are shown below in Table 4.

Table 2 Savings of distance and time for each trip

	Distance saved (in metres)	Time reduced (in minutes)
Upper route planner software	860	23
Field trial 1	640	16
Field trial 2	645	20
Field trial 3	650	17
Field trial 4	655	14
Field trial 5	660	15
Field trial 6	665	16
Field trial 7	670	9

Table 3 Savings

Type of savings	Amount saved
Fuel savings	Rs. 1015
Maintenance cost	Rs. 139
Labor cost	Rs. 994
Total cost saved per vehicle	Rs. 20,553

Table 4 Results

	Distance(in kms)	Time
Actual data	14.15	3 hr 13 mins
Upper route planner	13.29	2 hr 50 mins
VROOM simulation	11.9	2 hr 43 mins
FlexSim simulation	12.17	2 hr 33 mins

3 Conclusion

Vehicle Routing Problem for LPG Cylinder distribution is considered and an LPP model is proposed to determine the minimum transportation distance. The usage of the optimization software has proved to be effective in reducing the reported overall operating cost of the organization. The results produced are yet to be tested using mathematical models. A study on the LPG cylinder distribution is carried out and an LPP mathematical model is formulated, the routes were optimized using the software mentioned and results were tested in real time. Table 4 indicates the results obtained. The major findings of the work are as follows:

- Finding an optimized route can have a huge impact on the overall business performance of the distribution center.
- Framing an LPP to attain an optimized route.
- Simulating the optimized route can give an easier understanding of the delivery network to the delivery agents.

- The actual distance of 14.15 kms is optimized to 13.29 kms.
- The time taken for the truck before optimization is 3 hr 13 mins and it is reduced to 2 hr 50 mins after optimization

References

1. Yuliza E, Puspita FM (2019) Branch and cut method for solving capacitated vehicle routing problem(CVRP) model of LPG gas distribution routes. Sci Technol Indones 4:105–108
2. Garside AK, Laili NR (2019) A cluster-first route-second heuristic approach to solve periodic multi-trip vehicle routing problem. Jurnal Teknik Industri 20:172–181
3. Lopes C, Correia A, Costa e Silva E, Monterio M, Borges Lopes R (2020) Inventory models with reverse logistics for assets acquisition in a liquefied petroleum gas company. J Math Ind. MSC: 90B05; 90B06; 90B50
4. Sangaiah AK, Tirkolaee EB, Goli A, Dehnavi-Arani S (2019) Robust optimization and mixed-integer linear programming model for LNG supply chain planning problem, vol 24. Springer-Verlag GmbH Germany, pp 7885−7905
5. Vinay VP, Sridharan R (2012) Taguchi method for parameter design in ACO algorithm for distribution–allocation in a two-stage supply chain. Int J Adv Manuf Technol 64(9–12):1333–1343
6. Umarani R, Selvi V (2010) Comparative analysis of ant colony and particle swarm optimization techniques. Int J Comput Appl 5(4):1–6
7. Sayyah M, Larki H, Yousefikhoshbakht M (2016) Solving the vehicle routing problem with simultaneous pickup and delivery by an effective ant colony optimization. J Ind Eng Manag Stud 3(1):15–38
8. Calvete HI, Gale C, Oliveros MJ, Valverde BS (2007) A goal programming approach to vehicle routing problems with soft time windows. Eur J Oper Res 177:1720–1733
9. Cetin S, Gencer C (2015) A heuristic algorithm for vehicle routing problems with simultaneous pick-up and delivery and hard time windows. Open J Soc Sci 3:35–41
10. Chen JF, Wu TH (2005) Vehicle routing problem with simultaneous deliveries and pickups. J Oper Res Soc 57:579–587

Evaluation of IoT Adoption Barriers in the Sustainable Indian Healthcare Supply Chain: A Pythagorean Fuzzy DEMATEL Approach

Mangesh Joshi

Abstract IoT adoption is still in its infancy in developing nations like India, despite considerable policy interest and the tremendous potential benefits it offers the healthcare supply chain. The purpose of this article is to identify and examine different barriers that could prevent the healthcare sector from using IoT. Fourteen significant barriers to IoT adoption were identified during the literature review phase. Some barriers were identified during thinking sessions with experts from various fields. The barriers are divided into cause and effect components using the Pythagorean fuzzy DEMATEL technique. As a conclusion, six factors belong to the impact group and eight elements belong to the cause group. 'High implementation and operating cost' is identified as the important barrier in the effect category and 'Complex architecture' is identified as an important barrier in the cause category as per (D + R) values. 'Lack of awareness about lot benefits' is identified as a non-influencing and not affecting barrier. Findings will assist decision-makers in comprehending the causes and effects of IoT acceptance.

Nomenclature

IoT	Internet of Things
DEMATEL	Decision-Making Trial and Evaluation Laboratory
PF DEMATEL	Pythagorean fuzzy DEMATEL
SC	Supply Chain
PFN	Pythagorean fuzzy numbers
TOPSIS	Technique for order performance by similarity to the ideal solution
DRM	Direct Relation Matrix
EPFDRM	Expected Pythagorean Fuzzy Direct Relation Matrix

M. Joshi (✉)
Department of Industrial/Mechanical Engineering, Shri Ramdeobaba College of Engineering and Management, Nagpur 440013, India
e-mail: joshimp@rknec.edu

A. Sachdeva et al. (eds.), *Recent Advances in Operations Management and Optimization*, Lecture Notes in Mechanical Engineering,
https://doi.org/10.1007/978-981-99-7445-0_9

PFTRM	Pythagorean Fuzzy Total Relation Matrix
TrPFN	Trapezoidal Pythagorean Fuzzy Number

1 Introduction

In the era of advancing technology, the Internet of Things (IoT) has emerged as a transformative force, revolutionizing various industries, including healthcare. In particular, the Indian healthcare supply chain stands to benefit significantly from the integration of IoT solutions, promising enhanced efficiency, traceability, and sustainability. However, the successful adoption of IoT in this context faces various barriers that need to be identified and understood to ensure a seamless implementation. The government has taken steps to improve the efficiency of the healthcare supply chain by introducing initiatives such as the National Health Mission and the National Pharmaceutical Pricing Authority. These initiatives have helped to reduce the cost of healthcare and improve access to quality healthcare services. Additionally, the government has also implemented various policies to ensure the availability of essential medicines and medical devices at affordable prices.

In order to increase productivity, visibility, and viability, the Internet of Things (IoT) is a technology made up of several electronic and smart gadgets [2]. IoT refers to the ability to integrate physical objects with the internet in order to accomplish certain valuable goals [12, 26]. IoT makes it possible for enterprises to securely communicate information about their products, materials, and services across the worldwide supply chain [14]. The implementation of IoT in supply chain and operations brings numerous benefits, including reduced risks and costs, enhanced resource efficiency, increased stakeholder engagement, improved revenue planning, optimized asset utilization, and the promotion of sustainability within the supply chain [4]. According to Umar et al. [22], it also offers perceptibility, litheness, and transparency across supply chain progressions and empowers businesses to meet the difficulties of sustainable SC.

The DEMATEL approach developed by Gabus and Fontela [7], establishes the interdependencies and linkages between the criteria. DEMATEL offers a causal-effect map to illustrate how the criteria are related to one another and how they impact one another [24]. To find logical linkages and relationships with direct effects, it can examine all relationships between sets of data. In healthcare supply chain management, we frequently require a sizable number of criteria for evaluating opportunities and challenges in IoT adoption. These criteria may also include erroneous and conflicting data. In order to address an MCDM problem, [27] modified the TOPSIS approach with PFS and took into account Pythagorean fuzzy numbers (PFN). To achieve the best outcome, they specified a PFN distance metric for the TOPSIS approach. Garg [16] used Einstein's operations which he then used to solve an issue involving decision-making. It can be challenging to determine criteria precisely in an uncertain setting. However, when making decisions in an uncertain setting, the

Pythagorean fuzzy set can be used efficiently. This interval is likewise a fuzzy set because its boundary is uncertain. Researchers frequently employ fuzzy numbers that are triangular and trapezoidal among other shapes [6]. DEMATEL approach based on Pythagorean fuzzy sets can therefore be thought of as an appropriate instrument to manage software quality evaluation [3].

This article is divided into the following sections. The literature for barrier identification to IoT adoption is reviewed briefly in Sect. 2. Fundamentals of Pythagorean fuzzy sets and Pythagorean fuzzy DEMATEL approach are presented in Sect. 3. The numerical outcomes after application of DEMATEL are discussed in Sect. 4. Finally, Sect. 5's concluding observations bring the paper to a close.

2 Literature Review

The literature review is done for barrier identification. By looking for numerous papers in journals that are indexed by Scopus, Web of Science, Google Scholar etc., different factors were found. IoT design, IoT factors, IoT implementation, IoT barriers, and other similar terms are among the search terms chosen. From the literature review, the list of barriers contributing to IoT adoption was prepared. IoT practitioners and designers also participated in parallel discussion sessions.

During the discussion, this research shortlisted fourteen pertinent barriers to IoT adoption. The relative importance/relevance of design requirements, the ease of accessing information, studies done for that factor, etc. were among the criteria used in the factor shortlisting process. Eight experts participated in the discussion. Four academic specialists and four business professionals make up the team of experts. Table 1 includes the list of fourteen pertinent barriers.

3 Materials and Methods

One of the established approaches for real-world decision-making issues that could involve ambiguity and uncertainty is the DEMATEL method. Fuzzy sets and their various variants are useful in decision-making situations when there is uncertainty. The DEMATEL approach has been created by researchers in intuitive and uncertain settings. Wu & Lee [25] investigated the DEMATEL method for building the skills of a global manager in a fuzzy environment. Their findings grouped competencies into cause-and-effect categories. The methodology followed for this study is shown in Fig. 1.

Table 1 Details of identified IoT adoption barriers

Classification	IoT barriers	Description	References
Environmental	High energy consumption	As more data centres and IoT devices are used, the need for power will rise	Desingh [5]
	Legal and regulatory standards regarding e-waste generation	There is still a lack of openness around IoT regulatory norms. E-waste can be produced from outdated IoT equipment and machinery	Sebastian and Gupta [19]
Economic	High implementation and operating cost	The implementation of IoT necessitates high-tech infrastructure and calls for management, employee, and supplier training, which affects businesses with high operational costs	Raj et al. [18]
	Technology affordability	Budget constraints might potentially be a barrier to IoT adoption	Majumdar et al. [13]
Technological	Complex architecture	Due to so many devices being connected across the web, the complexity of architecture creates problems	Kamble et al. [11]
	Lack of knowledge management system	Lack of a standard knowledge management system that allows for end-to-end visibility	Ani et al. [1]
	Data management difficulty	One of the major challenges associated with big data is effectively storing the vast volumes of information generated. Companies' data centres and databases are constantly accumulating an ever-expanding amount of data, making it increasingly difficult to manage and handle these rapidly growing datasets over time	Zhong et al. [28]
	Lacking IoT suppliers & service providers	The IoT is an innovative and rapidly evolving platform that has gained significant traction in recent times. Identifying new suppliers and service providers, especially in emerging nations, presents challenges due to the rapid advancements in technology and frequent shifts in business strategies	Valmohammadi et al. [23]
	Lacking internet coverage and IT infrastructure	IoT adoption is hampered by poor internet availability and electrical issues	Sun et al. [20]

(continued)

Table 1 (continued)

Classification	IoT barriers	Description	References
	Lacking security & privacy concerns	It makes sense that systems can experience a variety of assaults. Data leakage and security vulnerabilities might both worsen as a result of multi-tenancy	Mineraud et al. [15]
	Data heterogeneity	Each linked device is subjected to diverse operating conditions and retains various sorts of data in an IoT-enabled environment	Rajput and Singh [17]
Organizational	Lacking qualified employees	Employees' lack of technological expertise might prevent businesses from using IoT	Talavera et al. [21]
	Lacking awareness about IoT benefits	Insufficient understanding of the benefits of IoT poses a substantial barrier to its adoption	Haddud et al. [10]
	Standardization	For two-way communication and information exchange among environments, smart objects, and other systems, standardization is crucial. It guarantees the efficient integration of data and stakeholders	Pang et al. [16]

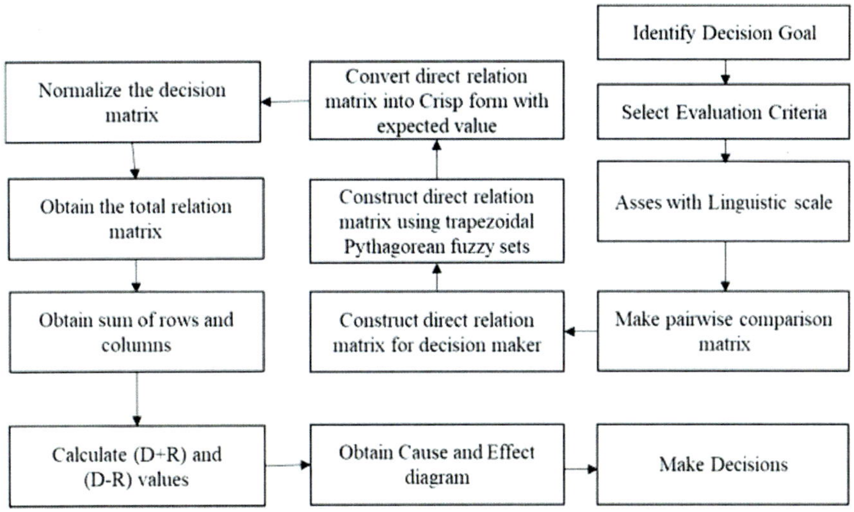

Fig. 1 Methodology

3.1 Pythagorean Fuzzy DEMATEL (PF DEMATEL) Method

Decision-makers typically base their judgment in decision-making on their experience and area of expertise. It can be challenging to accurately evaluate the measures for DEMATEL or other decision-making approaches in an uncertain setting. When making decisions, the Pythagorean fuzzy set successfully handles an uncertain environment. The Pythagorean fuzzy DEMATEL technique assesses assessment measures in terms of the TrPFN and the PF linguistic scale. We take into account a team of eleven experts that consists of three professors, five researchers, and three students who are employed in the area of our study that is relevant to them. Each expert's evaluation is used to gather linguistic information.

Step 1: Preparation of DRM based on the opinions of the experts.

Step 2: Normalize the EPFDRM.

Step 3: Construction of PFTRM.

Step 4: Generating a casual diagram.

In this section, the authors provide some introductions to PF sets and briefly explain the traditional DEMATEL approach, and Pythagorean fuzzy DEMATEL method.

4 Results and Discussions

The evaluation of IoT adoption barriers involves numerous complex variables. This study outlines a method for evaluating these barriers, which is based on a literature review. Through this review, 14 major criteria for IoT adoption barriers were identified and are presented in Table 1. To determine the relationship between these criteria, a linguistic analysis was conducted. A 7-point scale, explained by linguistics phrases and TrPFN was used to evaluate the criteria. The TrPFN equivalent predicted values were determined according to reference [9]. The results of this analysis are presented in Table 2, which shows the initial DRM.

The EPFDRM is generated by replacing the TrPFN with the appropriate expected value. To obtain the Normalized PF-DRM, each cell is divided by the maximum row summation value of the expected PFDRM. Table 3 represents PFTRM.

The sum of rows is stored in column D, and its sum of columns is in R. The values of (D + R) and (D-R) are then calculated. The (D + R) and (D-R) for each criterion are shown in Table 4. Figure 2 displays the causal diagram. This graphic shows the understanding of the significance of the IoT adoption barriers as the appreciation of the whole complicated structure.

From Fig. 2 and Table 4, it can be observed that the barriers to IoT adoption are divided into cause-and-effect groups based on the value of D + R and D-R. In this case, cause group embraces the barriers {B1, B2, B4, B5, B9, B10, B13, and B14} and the effect group includes the barriers {B3, B6, B7, B8, B11, B12}. Numerous other cues can be gleaned from Fig. 2.

Table 2 Initial direct relation matrix

Barriers	B 1	B 2	B 3	B 4	B 5	B 6	B 7	B 8	B 9	B 10	B 11	B 12	B 13	B 14
B 1	0	H	AH	AH	ML	FH	H	AH	L	AH	L	H	FH	H
B 2	AL	0	AH	AL	FH	AL	ML	AH	ML	ML	FL	L	FH	L
B 3	H	L	0	AH	ML	AH	ML	AH	ML	AH	AL	H	AH	H
B 4	AH	AL	H	0	FL	FH	AH	H	ML	H	FL	FH	H	FH
B 5	AH	FH	H	L	0	ML	FL	FH	FH	L	ML	FH	H	FL
B 6	ML	L	AH	L	FL	0	AH	ML	AL	ML	FH	FH	AH	H
B 7	FH	ML	FH	AL	L	ML	0	AH	ML	FH	ML	ML	FH	ML
B 8	L	AL	FH	ML	AL	AL	AH	0	H	AH	ML	FH	FL	ML
B 9	AH	ML	AH	ML	ML	FH	AH	H	0	ML	AL	H	FH	FH
B 10	H	ML	ML	FH	L	AH	AH	AH	H	0	ML	FH	ML	ML
B 11	L	AL	ML	AL	AL	AL	AL	FL	FH	ML	0	ML	ML	L
B 12	H	AL	H	AL	L	FH	AH	H	ML	L	ML	0	L	AH
B 13	AH	ML	AH	L	ML	H	AH	ML	FL	ML	ML	FH	0	AH
B 14	AH	L	H	ML	ML	AH	AH	H	L	ML	FL	ML	H	0

Table 3 PFTRM (T)

Barriers	B-1	B-2	B-3	B-4	B-5	B-6	B-7	B-8	B-9	B-10	B-11	B-12	B-13	B-14
B 1	0.28	0.18	0.42	0.27	0.13	0.30	0.43	0.45	0.18	0.37	0.13	0.30	0.30	0.32
B 2	0.15	0.05	0.27	0.09	0.11	0.13	0.22	0.29	0.13	0.17	0.08	0.13	0.18	0.14
B 3	0.37	0.12	0.32	0.27	0.14	0.36	0.41	0.44	0.19	0.37	0.12	0.30	0.36	0.33
B 4	0.35	0.09	0.35	0.14	0.11	0.26	0.42	0.37	0.17	0.30	0.12	0.25	0.29	0.27
B 5	0.31	0.13	0.30	0.13	0.07	0.20	0.26	0.28	0.16	0.19	0.11	0.21	0.25	0.20
B 6	0.24	0.09	0.34	0.13	0.09	0.18	0.36	0.28	0.11	0.22	0.14	0.21	0.29	0.26
B 7	0.21	0.10	0.24	0.10	0.07	0.17	0.21	0.31	0.13	0.20	0.10	0.17	0.19	0.18
B 8	0.19	0.06	0.24	0.13	0.06	0.15	0.32	0.20	0.18	0.26	0.10	0.19	0.16	0.18
B 9	0.36	0.13	0.39	0.18	0.12	0.27	0.42	0.38	0.14	0.26	0.10	0.28	0.27	0.27
B 10	0.31	0.12	0.30	0.18	0.09	0.30	0.41	0.39	0.21	0.21	0.13	0.24	0.24	0.24
B 11	0.10	0.03	0.13	0.05	0.03	0.07	0.10	0.12	0.10	0.11	0.03	0.10	0.10	0.09
B 12	0.27	0.07	0.29	0.11	0.08	0.22	0.35	0.31	0.14	0.19	0.11	0.15	0.18	0.28
B 13	0.36	0.13	0.39	0.16	0.12	0.29	0.41	0.33	0.15	0.26	0.13	0.25	0.22	0.33
B 14	0.34	0.10	0.35	0.17	0.12	0.31	0.40	0.35	0.14	0.25	0.12	0.22	0.29	0.21

The cause criteria group may have a significant influence on the global structure. Hence, the cause group criteria can receive more attention. 'Lack of qualified employees', 'Legal, regulatory standards', 'Availability of internet and IT facilities', 'High energy consumption, 'Complex architecture', 'Standardization', 'Data heterogeneity', and 'Technology affordability' are in cause group. In the cause group, 'Legal

Table 4 Segregating factors

Criteria	Sub-criteria	D	R	D + R	D-R	Identity
Environmental	High energy consumption	2.138	1.415	3.552	0.723	Cause
	Legal and regulatory standards	2.804	1.349	4.153	1.455	Cause
Economic	High implementation and operating cost	4.110	4.337	8.447	-0.227	Effect
	Technology affordability	3.370	3.365	6.735	0.005	Cause
Technological	Complex architecture	4.055	3.832	7.887	0.223	Cause
	Lacking KMS	2.962	3.212	6.174	-0.250	Effect
	Data management problems	2.385	4.723	7.109	-2.338	Effect
	Lacking IoT suppliers/service providers	2.420	4.476	6.896	-2.056	Effect
	Lacking internet coverage and IT facilities	3.570	2.128	5.697	1.442	Cause
	Lacking security concerns	2.750	2.987	5.736	-0.237	Effect
	Data heterogeneity	3.386	3.292	6.678	0.095	Cause
Organizational	Lack of qualified employees	3.483	2.124	5.607	1.358	Cause
	Lack of awareness about a lot of benefits	1.154	1.538	2.692	-0.384	Effect
	Standardization	3.524	3.332	6.855	0.192	Cause

and regulatory standards has the highest value of $(D - R)$, i.e., 1.455 which indicates that it impacts the entire structure more. Additionally, Table 4 demonstrates that the (D + R) of 'Complex architecture' is 7.887, demonstrating the significance as a barrier to IoT adoption. Complexity always tries to take away resources in the form of time, money, efforts, etc. Because of this, 'Complex architecture' is regarded as the most important metric for assessing IoT adoption challenges. The order of relative importance of the barrier based on (D + R) values is given in Table 4. 'Technology affordability' is having least influence on the entire system in the cause group.

In general, other criteria have a smooth impact on the effect group's criteria, rendering the effect group's criteria inappropriate for the evaluation. Out of the fourteen criteria considered in this research, six barriers, namely B3, B6, B7, B8, and

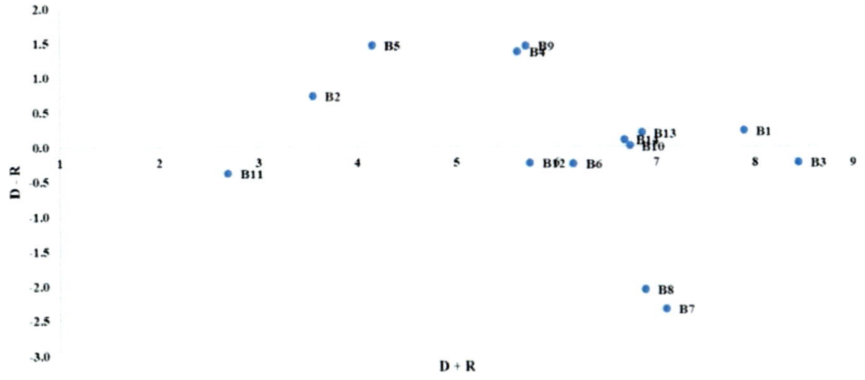

Fig. 2 Cause and effect diagram

B11, belong to the effect group. The criteria 'High implementation and operating cost' (B3) has the highest $(D + R)$ value of 8.447 which indicates that it is the most important barrier to IoT adoption in the effect group. As cost is always a concern of the organization, the implementation and operating costs are important barriers. 'Lack of awareness about lot benefits' has the lowest (D + R) value in the effect group and its (D-R) value is also very low, i.e., -0.384. Hence, this barrier may be considered a non-influential barrier. From Fig. 2, it is clear that 'Lack of awareness about lot benefits' (B11) is not influencing any of the other barriers. It is more independent in nature.

In conclusion, the Pythagorean fuzzy DEMATEL technique accurately identifies all these crucial barriers that are to be focused upon.

5 Conclusions

The implications drawn from the study carry significant relevance for multiple stakeholders involved in IoT development and implementation in the Indian healthcare supply chain. IoT developers and suppliers can utilize the study's findings to gain valuable insights into the factors influencing IoT adoption. By identifying the specific criteria impacting adoption, they can fine-tune their offerings and strategies to address the most critical barriers effectively. Project managers and implementers in the healthcare sector can leverage the study's results to assess the potential impact of different criteria on the successful integration of IoT solutions. Focusing on the cause-group criteria, which exert the most substantial influence on the effect-group criteria, allows them to prioritize their efforts and resources to overcome the primary obstacles. The research also contributes to a deeper understanding of the interconnectedness between various barriers to IoT adoption. By unraveling the relationships and dependencies among these obstacles, stakeholders can gain clarity on the complexity of the challenges they face. This enhanced understanding lays the foundation for the development and implementation of more effective strategies to promote IoT adoption in the healthcare supply chain.

As for future directions, the study suggests that applying its findings in real-life scenarios by collecting actual data will be a valuable step. This empirical approach can further validate the study's conclusions and provide real-world insights into the practical implications of the identified barriers. Furthermore, the scope of similar research can be extended to other related fields or industries to explore how IoT adoption barriers manifest in different contexts. By conducting similar studies in allied areas, researchers can identify common challenges and tailor solutions that are adaptable across various domains, thereby fostering broader IoT integration and advancement in different sectors.

References

1. Ani UD, He H, Tiwari A (2019) Human factor security: evaluating the cybersecurity capacity of the industrial workforce. J Syst Inf Technol 21(1):2–35
2. Awan U, Sroufe R, Shahbaz M (2021) Industry 4.0 and the circular economy: a literature review and recommendations for future research. Bus Strat Environ 30(4):2038–2060
3. Giri BC, Molla MU, Biswas P (2022) Pythagorean fuzzy DEMATEL method for supplier selection in sustainable supply chain management. Expert Syst Appl 193:116396
4. Belhadi A, Kamble S, Gunasekaran A, Mani V (2021) Analyzing the mediatingrole of organizational ambidexterity and digital business transformation on industry 4.0 capabilities and sustainable supply chain performance. Supply Chain Manag: Int J
5. Desingh V (2021) Internet of things adoption barriers in the Indian healthcare supplychain: an ISM-fuzzy MICMAC approach. Int J Health Plan Manag
6. Dubois D, Prade H (1983) Ranking fuzzy numbers in the setting of possibility theory. Inf Sci 30(3):183–224
7. Gabus A, Fontela E (1972) World problems, an invitation to further thought within the framework of DEMATEL. Battelle Geneva Research Centre, 1002 Switzerland, Geneva
8. Garg H (2016) A new generalized Pythagorean fuzzy information aggregation using Einstein operations and its application to decision making. Int J Intell Syst 31(9):886–920
9. Grzegrorzewski P (2003) The hamming distance between intuitionistic fuzzy sets. In: Proceedings of the 10th ifsa world congress. Istanbul, Turkey, pp 35–38
10. Haddud A, DeSouza A, Khare A, Lee H (2017) Examining potential benefits and challenges associated with the internet of things integration in supply chains. J Manuf Technol Manag 28(8):1055–1085
11. Kamble SS, Gunasekaran A, Sharma R (2018) Analysis of the driving and dependence power of barriers to adopt industry 4.0 in Indian manufacturing industry. Comput Ind 101:107–119
12. Kazancoglu Y, Ozbiltekin-Pala M, Sezer MD, Kumar A, Luthra S (2022) Circular dairy supply chain management through internet of things-enabled technologies. Environ Sci Pollut Res 1–13
13. Majumdar A, Garg H, Jain R (2021) Managing the barriers of Industry 4.0 adoption and implementation in textile and clothing industry: interpretive structuralmodel and triple helix framework. Comput Ind 125. Article 103372
14. Mishra D, Gunasekaran A, Childe SJ, Papadopoulos T, Dubey R, Wamba S (2016) Vision, applications and future challenges of internet of things: a bibliometric study of the recent literature. Ind Manag & Data Syst

15. Mineraud J, Mazhelis O, Su X, Tarkoma S (2016) A gap analysis of internet-of-things platforms. Comput Commun 89:5–16
16. Pang Z, Zheng L, Tian J, Kao-Walter S, Dubrova E, Chen Q (2015) Design of aterminal solution for integration of in-home health care devices and servicestowards the internet-of-things. Enterprise Inf Syst 9(1):86–116
17. Rajput S, Singh SP (2018) Identifying industry 4.0 IoT enablers by integrated PCA-ISM-DEMATEL approach. Manag Decis 57(8):1784–1817
18. Raj A, Dwivedi G, Sharma A, de Sousa Jabbour ABL, Rajak S (2020) Barriersto the adoption of industry 4.0 technologies in the manufacturing sector: an inter country comparative perspective. Int J Prod Econ 224. Article 107546
19. Sebastian VJ, Gupta R (2018) Retail ecosystem in India—an overview of the regulatory framework and the emerging paradigm. Theor Econ Lett 08(03):183–196
20. Sun QB, Liu J, Li S, Fan CX, Sun JJ (2010) Internet of things: summarize onconcepts, architecture and key technology problem. J Beijing Univ Posts Tele-commun 3(3):1–9
21. Talavera JM, Tobón LE, Gómez JA, κ.ά (2017) Review of IoT applications in agro-industrial and environmental fields. Comput Electron Agric 142:283–297
22. Umar M, Khan SAR, Yusliza MY, Ali S, Yu Z (2021) Industry 4.0 and greensupply chain practices: an empirical study. Int J Prod Perform Manag
23. Valmohammadi C, Ebrahimi F, Mohammadi M (2017) Proposing a model to study the impact of RFID technology on organizational performance. Libr Rev 66(1/2):69–82
24. Wu HH, Tsai YN (2011) A DEMATEL method to evaluate the causal relations among the criteria in auto spare parts industry. Appl Math Comput 218:2334–2342
25. Wu WW, Lee YT (2007) Developing global managers competencies using the fuzzy DEMATEL method. Expert Syst Appl 32(2):499–507
26. Wortmann F, Flüchter K (2015) Internet of things. Bus & Inf Syst Eng 57(3):221–224
27. Zhang X, Xu Z (2014) Extension of TOPSIS to multiple criteria decision making with Pythagorean fuzzy sets. Int J Intell Syst 29(12):1061–1078
28. Zhong RY, Newman ST, Huang GQ, Lan S (2016) Big data for supply chain management in the service and manufacturing sectors: challenges, opportunities, and future perspectives. Comput Ind Eng 101:572–591

A Framework to Implement Green Supply Chain Management for Sustainable Development

Vijay Sharma and Arvind Bhardwaj

Abstract Environmental sustainability is of utmost concern for almost every industry and government in the present scenario. This paper aims to suggest a framework for green supply chain management (GSCM) implementation in the dairy and agro industries in order to achieve sustainability and competitiveness. It has been revealed that GSCM is an emerging new and effective approach to improve productivity, efficiency and to achieve sustainability. The scope of the research work has been on the dairy and agro-industries of northern India and almost 92 dairy and 140 agro-industries have been examined for the purpose. The research methods employed are a survey through a detailed questionnaire, semi-structured interviews of senior officials, e.g., purchase managers, production managers, quality assurance managers, senior chemists as well as lead auditors of ISO 9000 and 14,001 system standards. A theoretical framework is derived from the literature to guide the research work. The intent of the framework is to have a systematic approach towards the implementation of green practices in an organization willing to adopt the GSCM approach. The outline of the framework is broadly divided into four phases, i.e**., plan-do-check-act** strategy. These four stages include all the green practices for effective implementation of green supply chain management. The framework provides the management a path to plan their green practices so as to achieve continuous improvement in environmental performance.

Keywords Green supply chain management (GSCM) · International organization for standardization (ISO) · Plan-do-check-act (PDCA) · Performance indicators (PI) · Chlorofluorocarbon (CFC)

V. Sharma (✉)
Mehrchand Technical Institute, Jalandhar, India
e-mail: er.vijaykumar_sharma@ymail.com

A. Bhardwaj
Dr. B. R. Ambedkar, National Institute of Technology, Jalandhar, India
e-mail: bhardwaja@nitj.ac.in

A. Sachdeva et al. (eds.), *Recent Advances in Operations Management and Optimization*, Lecture Notes in Mechanical Engineering,
https://doi.org/10.1007/978-981-99-7445-0_10

1 Introduction

The concern of the government along with the private sector has grown multifold towards environmental sustainability in the recent past to green their supply chains. Green supply chain management (GSCM) is a system improvement approach. Many industries have shown interest in implementing more and more green practices in order to have a cutting edge over their competitors. In this approach, a green factor is incorporated into the traditional supply chain by embracing green practices, e.g., green design, green manufacturing, green purchasing, green transportation etc. [1] suggested that in order to have a cutting edge over your competitors, one should think of adopting the GSCM concept. By seeing the environmental guidelines and rising demands by foreign buyers for a lot of eco-friendly products, many organizations across the sphere are embracing environment-friendly products as their professional strategy for future sustainability and to have a competitive advantage over others. In many countries, the Government, industries and some private organizations are operating along to purchase eco products which would finally advantage the environment and thus the society. Ho et al. [2] described that environmental management strategies should be developed due to changed environmental needs, further stating that green supply chains restrict the waste within the industry so as to save energy and the environment. Restraining damage to the surroundings has been the major challenge civilization faces over the coming years. Companies are gradually being advised to play their part in taking action to prevent long-term irreparable damage to our planet and they are scheduling their environmental programs to include green policy throughout their various stages of the supply chain. GSCM targets to control wastes within the production so as to conserve energy and prevent the dissipation of harmful material into the environment. Organizations are taking steps to develop into more eco-mindful and going green within the limitations of their trade intents. GSCM has emerged as an innovative concept to green the existing supply chain management. The present study focuses on the framework for the implementation of GSCM in dairy and agro companies and provides them with essential performance indicators and sub-indicators in order of their dominance. The introduction of GSCM in agro industry will improve their product design, procurement procedure, internal processing, distribution and re-processing operations, etc.

2 Research Context

The purpose of this paper is to suggest a framework for the implementation of green supply chain management in the dairy and agro-industries and to establish some green index for the evaluation of green practices in these companies. The Indian dairy and agro industries have been focused on for the study. The dairy and agro industries are selected due to the following reasons:

- These industries are major energy users.

- Due to increased incidents of food contamination, the dairy and agro industries are selected since food is the prime concern for the humankind. Moreover, dairy and agro products are of daily use and directly affect the health of a person. Dairy and agro industry also contribute almost twenty percent of the GDP of the Indian economy.
- Dairy and agro industrial sectors also accounted for more than fifty percent of the total workforce in India. India is the largest milk producer in the world to date but negligible work has been done on greening the dairy and agro sectors.

3 Literature Review

According to [3], GSCM can be defined as "incorporating environmental philosophy into supply chain management, including product design, material obtaining and selection, manufacturing method, delivery of the final product to the customers as well as end-of-life management of the product after its valuable lifespan". Hsu [4] define GSCM as an "enthusiastic viewpoint for the improvement in products & processes in respect of environmental performance within the boundaries of environmental regulations".

Olugu [5] define it as "GSCM eliminates or minimize based in the form of energy, emission, hazardous chemical and solid waste. Deif et al. [6] emphasizs system model approach for green manufacturing, which involves the design, planning and control of green practices related to manufacturing. Results are further validated through a case study. System modeling followed by a case study has been used as methodology. Baines et al. [7], the paper highlights the importance of green strategies in the context of new business opportunities and product development. It states that with the increasing consciousness of environmental issues, the customer is demanding more and more green products and the companies having green strategies will be the leaders in the coming future. Mosoumik et al. [8], carried out a study on Malaysian manufacturers see the impact of three green strategies, e.g. pollution prevention, product stewardship and clean technology. The results show that there is a need to promote product stewardship and clean technology whereas pollution prevention is already given considerable attention. Importance-performance matrix analysis (IPMA) has been used for analysis [9]. The purpose of this paper is to rank the key performance indicators that are responsible for the implementation of GSCM in the Indian dairy sector. Zhu [10], explored the economic model to find out the green product development problems with different supply chain structures. The results also look into government regulators and green NGO that promote green policies. The game theoretic approach has been used for modelling which starts with one manufacturer and one retailer, i.e. simple supply chain. Sharma et al. [11] the study is focused on the evaluation of the Indian dairy and agro industries on the green performance index in order to effectively implement GSCM. Vijay Sharma et al. [12] this paper is deliberated on finding green supply chain management related performance indicators in the Indian agro industry. The study suggests 13 performance indicators and 79 sub-indicators and ranks them in terms of their importance. Wong et al. [13] focus on green customer and supplier integration and collaboration to improve environmental performance and its inter alia benefits. Samad et al. [14] aim to investigate the alliance of the GSCM system and the performance of manufacturing companies using the Natural way as well as Institutional Theory.

4 Methodology

The research was initiated with an extensive literature review on GSCM, in-depth interviews with industrial experts and ISO auditors for the selection of the research topic. Thereafter, research problems were identified after a thorough gap analysis of the existing literature. The research aims and objectives were derived after several revisions of the literature. To achieve the derived aims and objectives, various issues related to GSCM implementation were studied in depth to develop some framework for implementing it. The main focus of the literature review was limited to green supply chain management in the dairy and agro industries. Two experts were from the academic side dealing with supply chain management and two experts from the industries dealing with the implementation of environmental management system (ISO 14000) and ISO 9000 etc. The questionnaires were mailed to the top and middle-level managers of purchase, production, quality control and testing laboratory departments in dairy and agro industries to collect the primary data. The data has also been collected through the process of semi-structured interviews. This data obtained is further used to construct a framework for GSCM implementation in the dairy and agro industries. Finally, the framework as shown in Fig. 1 is suggested based on plan-do-check-act (PDCA) approach. This framework is unique and very helpful in implementing GSCM in dairy and agro industries.

5 Framework for GSCM Implementation in Dairy and Agro Industries

5.1 Introduction

The intent of the framework is to have a systematic approach towards the implementation of green practices in an organization willing to adopt the GSCM approach. The outline of the framework is broadly divided into four phases, i.e., plan-do-check-act strategy. These four stages include all the green practices for effective implementation of green supply chain management. The framework provides the management a path to plan their green practices so as to achieve continuous improvement in environmental performance. These four stages of implementation are explained hereunder.

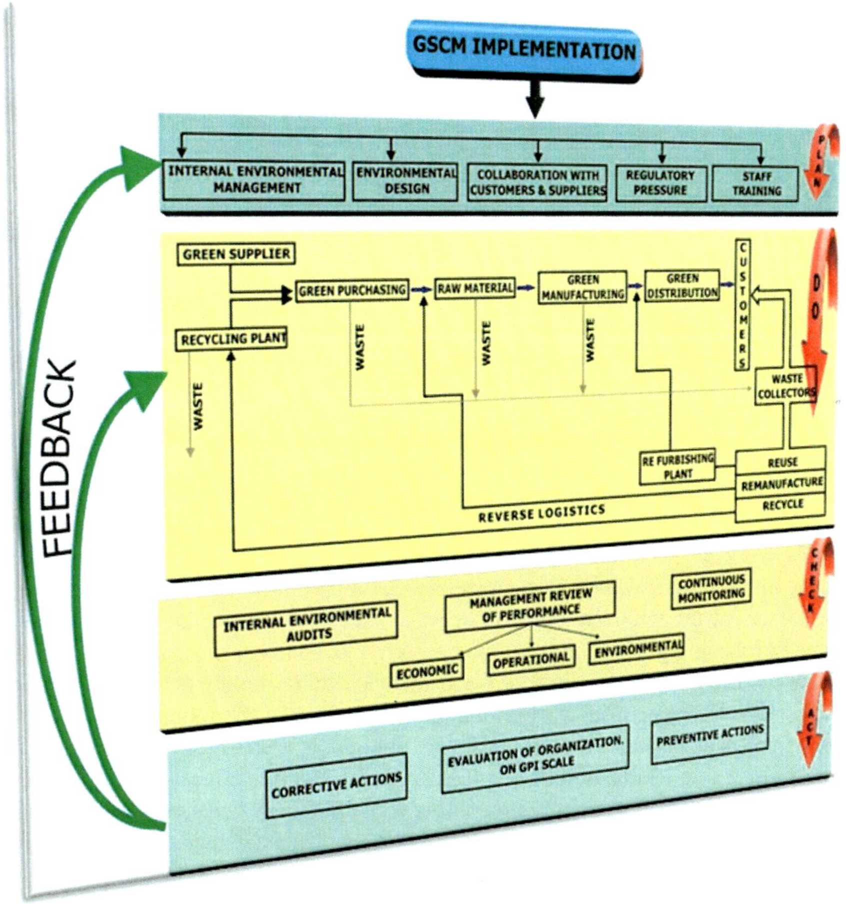

Fig. 1 Framework for GSCM implementation

5.2 *Plan Phase*

This phase encompasses five elements namely internal environmental management, environmental design, collaboration with customers and suppliers, regulatory pressures and staff training. In internal environmental management, the top-level management described the company's environmental policy to ensure environmental protection, and to comply with the Government regulations related to the environment and its objectives. It reflects the commitment of top-level management. The environmental policy should include the essence of continuous improvement and sustainable development, minimization of waste throughout the supply chain and pledge to adopt recycling, reuse, remanufacturing, etc. Green design is a vital part of the product lifecycle since most of the environmental concerns are taken into consideration during this phase. The product material, processes, its packaging material, etc. can be designed in such a way as to reduce energy consumption or to control the wastage during the manufacturing stage. The product can be designed with a view to recycle, reuse or re-furbish after the end of the useful life of the product. The use of less polluting and less hazardous materials can also be planned during this phase. Collaboration with customers and suppliers is considered in the planning stage for the successful implementation of GSCM throughout the supply chain. If the company focuses on its in-house activities only and not on suppliers and customers, then the whole holly purpose of protecting the environment would be forfeited.

5.3 *Do-Phase*

This phase assimilates environmental criteria into the whole supply chain. It restricts all the waste within the production cycle so as to reduce its harmful impacts on the environment. Conventional supply chain hardly takes care of their deleterious effects on the environment whereas green supply chain takes into consideration the negative effects of the whole production cycle on the natural environment. It starts with green purchasing where green supplier selection is done to procure the eco-friendly materials. Environmental objectives are shared with the suppliers and development of suppliers is carried out during this phase through vendor rating based on environmental performance. Development of suppliers is done by keeping in mind the long-term relationships. Purchasing the green raw material reduces its adverse effect on the environment. Green procurement is very effective when suppliers have long-term policies in implementing green supply chain management. In conventional procurement, the prime focus is always on the price but in green procurement, the ecological factor is the main criterion for supplier selections. Green manufacturing aims to lessen the negative impact of the production processes on the environment. It initiates with green design which involves the green factor during the product design to reduce its negative impact on the environment. The effect of the product on the environment is determined during this phase. All operational and technical aspects of the product are closely observed in this phase. A well-designed product by taking into account environmental considerations reduces its ill effects on the environment throughout its life span. Green design is associated with the product or its component which makes that product. It also involves packaging or manufacturing processes to reduce the consumption of energy. The green design also reduces production waste and maximizes the profit of the company by recovering material and reusing, recycling, or remanufacturing it. The green design mainly includes recyclable, less hazardous, and non-polluting material processed through energy-efficient processes. In green manufacturing phase, it is recommended to consume less energy and raw materials and to reduce waste. It should also reduce harmful emissions and the use of toxic material. Better and energy-efficient production processes should be used to lower the manufacturing cost. Since waste and reworking result in extra material consumption and energy reducing waste and optimizing production methods are the main activities of green manufacturing. Most of the companies surveyed had ISO 9000 or ISO 14001 (Environmental management system) systems so it would be easier to implement GSCM in these companies since by adopting ISO standards quality-conscious and environment-conscious atmosphere is created in these companies. Green distribution should adopt sustainable and less polluting transportation systems because it is one of the main causes of greenhouse gas release. To lessen the effect of transport on the environment combined transportation, e.g., rail-road, sea-road, road-air can be used. Especially in dairy and agro industries, the transport vehicle should use eco-friendly gases like hydrocarbon (HC) instead of chlorofluorocarbon (CFC). Collaboration with other similar industries should also be encouraged to optimize the use of such vehicles. In reverse logistics, the dairy and agro companies will receive back the packaging material in order to recycle or reverse management practice wherein the material is recovered back after the end of the useful life of the product. The reverse logistic is encouraged due to Government environmental regulations, the severe competition which forces these industries to be environmentally friendly and for the purpose of recycling or reusing it. The reverse logistic system can be built up by these companies individually or by a group of similar companies or it can be outsourced to a third party. The reverse logistics greatly depends on the design of the product. If the product is green-designed by taking into account the reuse or recycling of material then it becomes easier to execute this exercise. Consumer behaviour actually governs the reverse logistic system of any company. The demand of the customer to use recyclable material is the deciding factor for the industry and compels the companies to establish to system of recovering the returned material from the consumers. Therefore, environmental performance, social obligations, economic factor and regulatory requirements are the main factors that determine the selection of a reverse logistic system.

5.4 Check-Phase

This phase includes internal environmental audits, management review and continuous monitoring. For the purpose of implementing GSCM, a company must have a documented procedure of an effective audit system and monitoring system.

5.4.1 Internal Environmental Audits

The company must have documented procedures to carry out environment audits at regular intervals to ensure whether or not the established system conforms to GSCM requirements.

- The audit system should be so established that it should provide an outcome of audit, i.e., audit reports to the top-level management.

- Responsibility and authority to conduct such audits should be clearly defined.
- The results of the previous audit should also be reflected in the audit reports.
- Procedure: For compliance with environmental procedures in order to implement GSCM internal and external audits are mandatory. These audits should be performed by a team of experts in the field of environmental regulations, engineering and auditing. The auditors should have got proper training in auditing skills or be qualified for ISO 14001 (Environmental management system). There should be a set procedure to conduct an audit and it includes a fixed format for audit reflecting designation and qualification of auditors, scope of audit, various activities to be audited, frequency of audits, etc. The internal audit can be carried out by the internal auditor, i.e., middle-level managers trained on internal audit skills. Team should have experts in different areas including environmental regulation, energy management, engineering and auditing skills. External audits are always more effective since they provide the neutral opinion of the auditors and provide a clear picture of the system to make better decisions. Internal staff may not be informed about the factual position which is very crucial for implementing the desired system, but this does not happen in third-party audits. Inter-departmental audits should also be conducted along with external audits so as to make timely decisions that can be fruitful for external auditing.

5.4.2 Management Review

Management review is a very important part of the GSCM implementable process. It ensures the continuous improvement of the organization towards its objectives to establish GSCM. Management review is always documented and aids the top management in evaluating the compliance with the system and its effectiveness in the light of the results of the review. Top management also suggests the possible need for change in environmental policy or its objectives.

Procedure: The top-level management must assign one management representative who is actually a leader of the green supply chain management implementations team. Management representatives will keep a record of all the internal as well as external audits and also the corrective and preventive actions to implement the GSCM system successfully.

- M.R. will also prepare the agenda for management reviews in consultations with senior officials of various departments. The results of the review are further used to make any amendments to the objectives and environmental policy of the company.
- The scope of management review should include economical, operational along with environmental factors into consideration.
- Management reviews are actually a continuous process and are not limited to framing environmental policy/objectives and implementation of GSCM.

The review should be documented showing the level of implementation of the GSCM system.

- Review should include compliance with regulatory environmental requirements, policy for waste reductions, etc.
- Determine the effects of manufacturing processes, energy conservation and internal and external audits on environmental performance.
- Finally corrective and preventive actions are suggested by the top management.

5.4.3 Continuous Monitoring

It is a routine checking or measuring of the crucial activities that lead to the implementation of green supply chain management. The key performance of GSCM indicators is regularly monitored and measured and recorded. The organization should have a definite procedure in the form of 'work instructions' to perform any activity. Monitoring ensures the compliance of environmental policy and objectives. Daily monitoring and measurement provide the individuals with the scope of improvement and identify the areas where preventive or corrective actions could be taken.

Procedure:

- Pinpoint the areas that are critical in terms of GSCM implementation.
- Identified the activities which have an impact on the environment.
- Keep a record of daily measurements, e.g., quality control, production, distribution, purchasing, etc.
- Relevant records should be generated with the name of person responsible for it.
- All the measuring instruments or equipment should be calibrated at regular intervals of time and recorded.
- The non-conformance should be rectified by corrective actions.
- Monitoring and measurement should be in alignment with company policy and its targets.
- The adequate resources to achieve desired environmental targets should be available.

5.5 Act-Phase

Corrective and preventive action: The company should establish a documented procedure to take corrective and preventive action for any non-conformity or deviation from the environmental objectives resulting from internal and external audits, management reviews or continuous monitoring. The corrective actions are necessary to implement GSCM effectively. There should be a proper record of every corrective or preventive action taken.

6 Procedure

- The environmental policy of the company and other objectives should be shared with each and every employee of the organization.
- It should be a problem-solving activity, not a fault-finding exercise. Everyone should be informed in the company that they can suggest positive changes aligning with company objectives.
- The responsibility and authority should be well defined so as to find out the actual cause of any non-conformance or failure.
- It is desired to have a separate team for recording the causes of action, corrective/ preventive actions taken, the effectiveness of corrective action taken, customer's complaints, waste reductions, etc.
- Proportionate actions should be taken in accordance with the level of problem/ non-conformance.
- Preventive actions should be initiated after a thorough analysis of the root cause of the problem. It is always taken to avoid similar problems in the future.
- In a nutshell, preventive actions find out the real cause of failure/non-conformity in process or other operational activity whereas corrective actions are taken to correct the deviations.

7 Conclusions

The rationale of this study was to suggest a framework for GSCM implementation to dairy and agro industries based in India. To achieve the objectives of the study, two case studies were conducted; one in dairy industries and the other in agro industries and the whole supply chain was studied in depth. Detailed interview of the officials at different levels and ISO 14001 and ISO 9000 auditors/ lead assessors were taken and their suggestions were noted down. Finally, a framework for GSCM implementation in Indian dairy and agro industries has been proposed. The framework also enlightens us as to how to do continuous improvement of the whole system to implement green supply chain management successfully. It also deliberates on the necessity of environmental policy and the role of management of the company to take corrective and preventive actions from time to time as suggested by the team of auditors.

8 Scope for Future Work

- The work can be extended further to compare the different types of industries in terms of GSCM implementations and evaluation of the green performance index.
- Future work can be focused on evaluating the effectiveness of the suggested framework for GSCM implementations in Indian dairy and agro industries.

- The research can be further carried out on the economic analysis of GSCM-adopted organizations to see the financial benefits of GSCM in terms of profit earning.
- The study can be concentrated on the identification of various barriers to GSCM implementations.
- Future recommendations for research work is to include customer and other entities, e.g., suppliers, distributors, retailers, etc. in case studies and data collection process.
- The study comparing GSCM and other environmental management systems and quality standards like ISO 9000 and ISO 14001 can be further directions of future research work.

References

1. Sarkis J (2011) An organizational theoretic review of GSCM literature. Int J Production Economics 130(1):1–15
2. Ho JC, Shalishali MK, Tseng TL, Ang DS (2009) Opportunities in green supply chain management. Coast Bus J 8(1)
3. Srivastava SK (2007) Green supply chain management: a state-of-the-art literature review. Int J Manag Reviews 9(1):53–80
4. Hsu W, Hu AH (2008) Green supply chain management in the electronic industry. Int J Environ Sci Tech 5(2):205–216
5. Olugu EU, Wong KY, Shaharoun AM (2010) A comprehensive approach in assessing the performance of an automobile closed-loop supply chain. Sustainability 2(4):871–889
6. Deif M (2011) A system model for green manufacturing. 6:27–36
7. Baines T, Brown S, Benedettini O, Ball P (2013) Examining green production and its role within the competitive strategy of manufacturers. J Ind Eng Manag 5(1):53–87
8. Masoumik SM, Abdul-rashid SH, Olugu EU (2015) Importance-performance analysis of green strategy adoption within the Malaysian manufacturing industry. Procedia CIRP 26:646–652
9. Sharma VK, Chandana P, Bhardwaj A (2015) Critical factors analysis and its ranking for implementation of GSCM in Indian dairy industry. J Manuf Technol Manag 26(6):911–922
10. Zhu W, He Y (2016) Green product design in supply chains under competition Wenge PT US CR. Eur J Oper Res
11. Sharma VK, Chandana P, Sachdeva A, Bhardwaj A (2016) Evaluation of Indian dairy and agro industry on green performance index for implementing green supply chain management. Vivechan Int J Res 7(1):1–10
12. Sharma V, Chandna P, Bhardwaj A (2017) Green supply chain management related performance indicators in agro industry: a review. J Clean Prod 141:1194–1208
13. Wong C, Wong C, Boon-itt S (2020) Effects of green supply chain integration and green innovation on environmental and cost performance. Int J Prod Res 58(15):4589–4609
14. Samad S, Nilashi M, Almulihi A (2021) Green supply chain management practices and impact on firm performance: the moderating effect of collaborative capability. Technol Soc 67:1017–1066

Convolutional Neural Network Based Image Processing Model for Supply Chain Management

Ashish Kumar, Saurabh Tiwari, and Sunil Agrawal

Abstract India follows China as the world's top fruit producer. A variety of factors faced at various stages along the supply chain cause 30–33% of the produced produce to be wasted annually. One of these factors is the storage and transportation of substandard fruit. In order to classify and grade fruits, this work intends to develop an efficient and highly accurate image processing model. To do this, we created a base Convolutional Neural Network (CNN) model and compared it with the modified pre-trained ResNet models. On the basis of the selected performance criterion, a thorough study of the models and comparison of them was conducted. All of the pre-trained ResNet models outperformed our base 3-layer CNN model, which had 83.8% accuracy, with Resnet18 and Resnet34 achieving the maximum accuracy of 97.30%. The created model can be incorporated into a real-time image processing system to guarantee that quality standards are adhered to across the whole supply chain.

Keywords Image processing · Convolutional neural networks · Transfer learning · ResNet

1 Introduction

Agriculture is one of the sectors that plays an important role in the economic development of emerging economies such as India. Producing fruits, vegetables, flowers, ornamental plants, spices, sauces, and beverages is the domain of horticulture, a

A. Kumar (✉) · S. Tiwari · S. Agrawal
Mechanical Engineering Department, PDPM IIITDM Jabalpur, Jabalpur 482005, India
e-mail: 1913603@iiitdmj.ac.in

S. Tiwari
e-mail: saurabhgtiwari@gmail.com

S. Agrawal
e-mail: sa@iiitdmj.ac.in

A. Sachdeva et al. (eds.), *Recent Advances in Operations Management and Optimization*, Lecture Notes in Mechanical Engineering,
https://doi.org/10.1007/978-981-99-7445-0_11

subfield of agriculture. The outlook for horticulture crops in India has improved greatly. Horticulture's contribution to agricultural output as a percentage in 2017–18 reached 33%. Among perishable items, such as fruits, India exported bananas, oranges, and apples worth ₹34,877.39 lacs, ₹3485.36 lacs, and ₹3916.59 lacs, respectively, during 2017–18 [1]. However, India's share in the global export market is still just around 1%, necessitating the use of real-time quality measuring system.

1.1 Research Background

Alarming food wastages and losses, growing concerns about the quality and safety of food, and weak economic stability are all features of perishable supply chains. The supply chain structure varies from place to place with respect to the perishable product but normally incorporates four stages namely producers, wholesalers, retailers, and customers. All these stages are linked to each other via storage and transport operations. In the supply chain of fruits, a major financial loss occurs due to the wastage of the products. These wastages are the consequences of a no. of factors, one of them being the storage and transport of substandard fruits that do not meet the quality requirements as expected by the consumer. Therefore, classification and grading of the fruits needed to be done. This can be done by humans but it is found to be inconsistent, time-consuming, variable, subjective, arduous, and expensive. Image processing can be used at different stages of the supply chain to reduce wastage and financial losses associated with it by correctly classifying and identifying the grades of the fruits, the details of which are mentioned below:

- Identification of substandard fruits: One category under substandard is rotten. Rotten fruits may rot the whole caret consisting of fresh ones also when stored with them. For instance, exposing a healthy apple to a rotten one will make the fruit ripen more quickly and eventually turn rotten. The explanation behind this is that as apples ripen, they release ethylene, a hormone that acts as a catalyst for fruit ripening. A well-ripened fruit's gas is absorbed by other fruits, causing them to ripen while the well-ripened fruit is nearby.
- Dynamic Pricing: Generally, transportation takes 2–5 days to reach their destination in a fruit supply chain. In India, most of the transportation tasks are carried out in non-refrigerated vehicles such as trucks. This results in exposing the fruits to highly unstable and undesirable environmental conditions. Also, the environmental conditions fluctuate drastically affecting the remaining shelf life of fruits and hence their quality. Therefore, image processing can be used before the transportation stage to grade the fruits and thereby change the prices associated with them. Also, substandard items that are not transported based on the trade-off between the estimated transportation days required and their remaining shelf life can be sold in the local market at reduced prices [2].

2 Previous Work

Most of the research work in the field of image processing using deep learning for fruit classification and grading tasks has been carried out in the previous demi-decade. Identification of the particular type of fruit seen in an image that contains a single type or numerous varieties of fruits is known as classification. Grading of fruits is one of the critical tasks to avoid losing added value in the markets. One of the crucial duties to prevent losing added value in the marketplaces is fruit grading. As a result, ongoing attempts are undertaken to enhance the techniques for determining fruit damage, infections, freshness, and maturity level. The three main application areas of image processing in the agricultural domain are classification [3], grading [4], and detection tasks [5].

The authors of the majority of the earlier research works [5–8] attempted to create CNN models for classification tasks. In the event that the accuracy attained was superior to constructed CNN models, they also implemented some pre-trained models. Using the Fruits-360 dataset, Sakib et al. [6] created and analyzed a number of CNN architectures for classifying fruits. They experimented with different combinations of hidden layers and epochs and achieved an accuracy of 100%. Lu et al. [7] developed CNN models and analysed the effect of data augmentation on the model's classifier's accuracy for classification tasks. It was found that incorporating data augmentation approaches improves the model's accuracy and leads to the development of a more robust model. The pre-trained model VGG utilised by Zeng et al. [8] provided better accuracy than the CNN models that they had constructed. The AlexNet, MobileNet, GooLeNet, VGGNet, and Xception architectures were compared against a 2-layer CNN model as a baseline for the grading task [9]. The performance of pre-trained models was compared to a baseline 2-layer CNN architecture by the authors. The results demonstrate that VGGNet gets the best accuracy for both datasets (96.49% and 89.12%), whereas GooLeNet was the most computationally efficient architecture by needing significantly less memory and training time.

2.1 *Objectives*

The development of an effective and highly accurate image-processing model that can classify and grade fruits is the main objective of the current work. In our research work, we are focusing on the classification and grading aspects of image processing. Also, based on the characteristics of the dataset, the grading of the fruits is carried out based on the freshness aspect only quantifies the quality of the product. The fruits are graded into two classes, i.e., fresh and rotten. The dataset consisted of images of three different classes of fruits namely bananas, apples and oranges and two classes of grades namely fresh and rotten ones for each one of them. To achieve our main objective, the different tasks performed are mentioned below:

- Development of novel base Convolutional Neural Network (CNN) model.

- Modification of pre-trained models of the ResNet family, i.e., Resnet18, Res net34, Resnet50, Resnet101, and Resnet152 using the concept of transfer learning.
- Comparison of the models, i.e., developed and modified in terms of accuracy, execution time as well as other model performance criteria like precision, recall, and F1 -score.

The remainder of the paper is structured as follows: Sect. 3 of the paper presents the research methodology. The model evaluation methods, findings, and discussion of the DL models are presented in Sect. 4. Section 5 concludes with conclusions and future research, and then lists references.

3 Research Methodology

Figure 1 displays the proposed methodology for creating image processing models together with its individual steps.

3.1 Data Collection and Preprocessing

Dataset to be used for image processing tasks generally possess two challenges namely insufficiency and imbalance both resulting in poor performance. In order to deal with above-mentioned challenges related to insufficiency, we selected a large dataset from the online platform Kaggle. The dataset included images of three different fruit classes and their two different fruit grades in the train and test folders. Now, to deal with imbalance problems, we eliminated low-sized images from all classes so that each class of the train dataset and each class of the test dataset has an equal no. of images. Each class in the test dataset has 1400 images while 300 images in the case of the test dataset resulting in an overall 10,200 images.

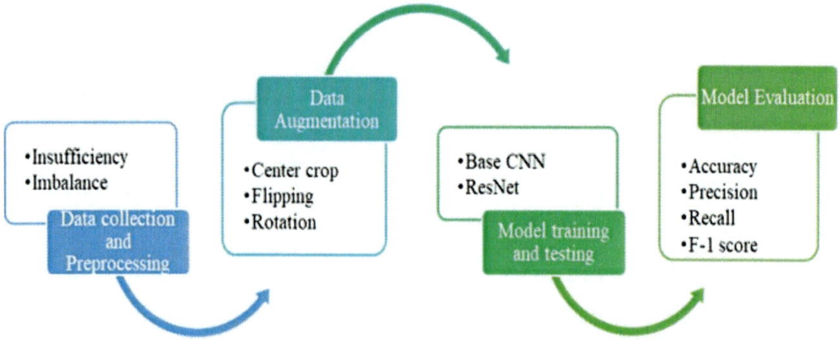

Fig. 1 Research methodology

Fig. 2 Random fruit images before (left) and after (right) data augmentation techniques

3.2 Data Augmentation

Image data augmentation is a technique for artificially increasing a dataset's size by producing altered versions of the dataset's images. More data can be used to train deep learning neural network models, which can lead to more accurate and efficient models. The various augmentation methods alter the original images, which helps fit models and apply what they have learned to new images more reliably [10]. In the research work, we performed center crop, random horizontal flip, random vertical flip and rotation on training images while only center crop was performed on test images. The description of performed data augmentation techniques is provided below:

- Center crop: We center cropped the images resulting in an image size of 224 × 224 pixels per inch, which is a pre-requisite for all pre-trained models.
- Flipping: We performed both random horizontal and random vertical flips with a probability of 0.5.
- Rotation: We rotated our dataset's images by 20 degrees.

Figure 2 shows some random images from each class of the training dataset before and after performing data augmentation techniques.

3.3 Model Training and Testing

Training and testing can be performed only after we have finalized our model's architecture. Next, the parameters of the model were finalized based on previous research work carried out on similar datasets [11] as well as by performing some initial experiments. In initial experiments, the effect of different parameters on our base CNN model's accuracy was analyzed. It was found that the majority of the parameters, including the activation function, optimizer function, and learning rate, exhibit effects that are similar to those indicated in earlier studies [11]. In the case of batch size selection, our model performed better with a batch size of 24 rather than 16. Also, we trained and tested for 20 epochs only and achieved similar accuracy. Based on collective information obtained from previous research work and our initial experiments, we finalized our model's parameters for all the models. Below are the descriptions of the common model parameters and their chosen values:

- Activation function: In all the models, we used the sigmoid activation function in the output layer.
- Optimizer function: In all the models, we used Adam as the optimizer function.
- Learning rate: In all the models, we used OneCycleLR scheduler and set the maximum learning rate of 0.01 during the initial 10 epochs and 0.001 during the final 10 epochs.
- Epochs: We trained and tested all our models for a total of 20 epochs.

Base CNN Model

In our base CNN model, we used 3 convolution layers, where in each of them, 3 operations were performed namely convolution, batch normalization followed by ReLU activation and Max pooling. The filter used in each convolution layer is of size 3 × 3. After this, two fully connected layers are added along with a dropout layer in between them. The first fully connected layer has 25,088 input features and 600 output features while the 2nd fully connected layer has 600 input features and 6 output features. Figure 3 describes the architecture of our base CNN model.

Transfer Learning Approach: ResNet

The goal of the machine learning technique known as transfer learning is to use knowledge obtained from an existing model to tackle a new problem. The present study modified the existing pre-trained CNN models of the ResNet family trained on

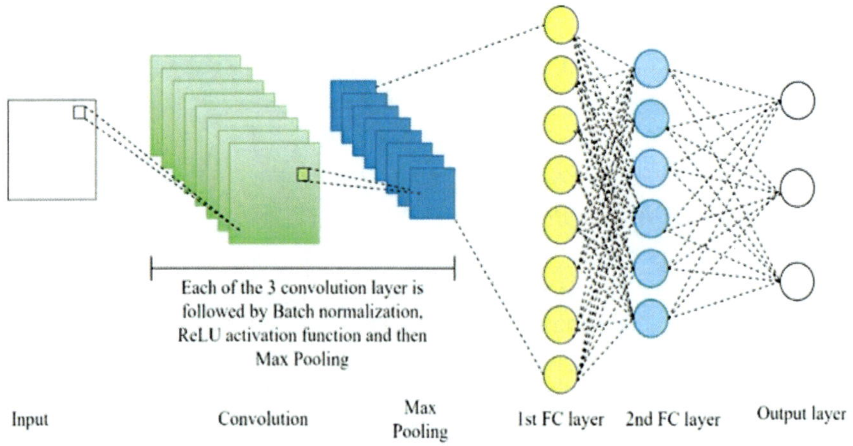

Fig. 3 Simplified diagram of our 3-layer base CNN model

the ImageNet dataset. In our paper, five pre-trained ResNet models are considered and evaluated for their performance on classification and grading tasks. The pre-trained models are Resnet18, Resnet34, Resnet50, Resnet101 and Resnet152. In order to use the pre-trained models, we modified the last fully connected layer of ResNet models to have the number of output classes equal to 6 as required based on our dataset.

3.4 Model Evaluation

A confusion matrix is a table that lists how many predictions a classifier made correctly and incorrectly. It is employed to evaluate a model's performance. The analysis of performance indicators like accuracy, precision, recall, and F1-score provided in the classification report may be used to assess the performance of a classification model. Also, the execution times of all the models are considered to determine the most efficient model.

$$Accuracy = TP + TN/(TP + TN + FP + FN) \tag{1}$$

$$Precision = TP/(TP + FP) \tag{2}$$

$$Recall = TP/(TP + FN) \tag{3}$$

$$F1 - score = (2 * Precision * Recall)/(Recall + Precision) \tag{4}$$

4 Results and Analysis

In this section, the result of all the models is shown and analysed with different performance criteria. Google Colab with Tesla K80 GPU and 16 GB memory was used to execute all the models. The confusion matrices and classification report of all the models were developed and based on them, precision, recall, and F1-score tables were created. The description of the model evaluation metrics and their analysis are:

Accuracy: Based on the accuracy versus epoch plot in Fig. 4, we reach the following conclusions (Table 1):

- The plot of the base CNN model unlike ResNet models shows most fluctuations and it achieved above 80% accuracy only in the last segment of epochs between 16 and 20. The fluctuations are observed because the base CNN model has to learn features of all the images from scratch and therefore, we see sudden spikes in the plot.

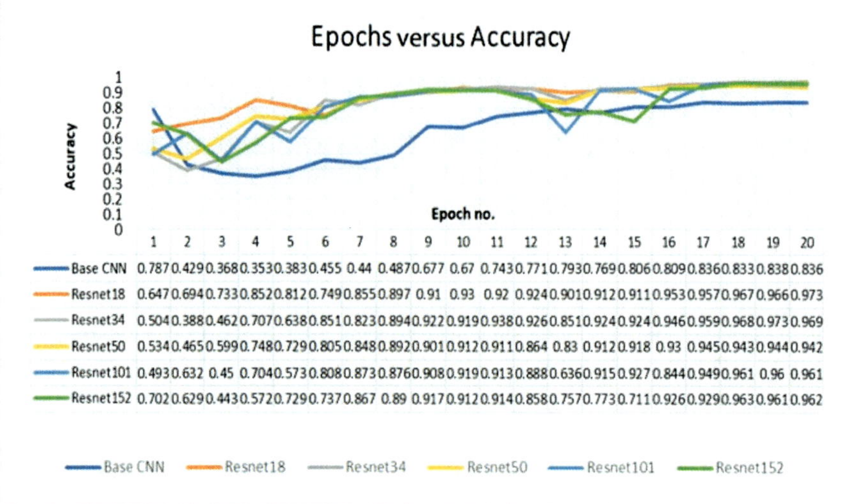

Fig. 4 Accuracy versus epoch plot

Table 1 Models and their execution times

Models	Base CNN	Resnet18	Resnet34	Resnet50	Resnet101	Resnet152
Execution time	61 min 4 s	37 min 49 s	39 min 9 s	56 min 7 s	63 min 3 s	214 min 2 s

- All the models except base CNN achieved more than 90% accuracy within 20 epochs. The highest accuracy of 97.3% is obtained by both Resnet18 and Resnet34 models in their 20th and 19th epochs, respectively.

Precision: Based on precision values mentioned in Table 2, we reach the following conclusions:

- Resnet18 model achieved the highest precision values for 2 classes named fresh apples and rotten apples while the Resnet 34 model achieved the highest precision values for 2 classes named fresh bananas and rotten bananas.

Table 2 Precision values of models

Models	Base CNN	Resnet18	Resnet34	Resnet50	Resnet101	Resnet152
Fresh apples	0.89	0.99	0.96	0.92	0.97	0.95
Fresh banana	0.89	0.96	0.99	0.96	0.95	0.95
Fresh oranges	0.90	0.97	0.98	0.97	0.98	0.98
Rotten apples	0.74	0.96	0.93	0.89	0.91	0.93
Rotten banana	0.84	0.99	1.00	0.97	0.99	0.99
Rotten oranges	0.75	0.97	0.97	0.93	0.96	0.97

- Resnet34, Resnet101, and Resnet152 perform equivalently well for the fresh oranges class while Resnet18, Resnet 34, and Resnet152 for the rotten oranges class.

Recall: Based on recall values mentioned in Table 3, we reach the following conclusions:

- Resnet18 model achieved the highest recall values for 2 classes named rotten apples and rotten bananas.
- Resnet34 and Resnet152 models performed equivalently well for the class of fresh apples.
- Resnet18, Resnet34, and Resnet50 performed equivalently well for the class of fresh bananas.
- Resnet18, Resnet34, and Resnet101 performed equivalently well for fresh oranges while Resnet101 performed best in the case of rotten oranges.

F1-score: Based on the F1-score values mentioned in Table 4, we reached the following conclusions:

- Resnet18 model achieved the highest F1-score values for 2 classes named fresh apples and rotten apples while the Resnet34 model achieved the highest value for the class named fresh banana.
- Resnet18, Resnet34, and Resnet101 performed equivalently well for fresh oranges while Resnet18, Resnet101, and Resnet152 in case of rotten oranges.

Table 3 Recall values of models

Models	Base CNN	Resnet18	Resnet34	Resnet50	Resnet101	Resnet152
Fresh apples	0.92	0.97	0.98	0.97	0.95	0.98
Fresh banana	0.94	1.00	1.00	1.00	0.99	0.99
Fresh oranges	0.80	0.98	0.98	0.93	0.98	0.97
Rotten apples	0.68	0.96	0.95	0.91	0.93	0.93
Rotten banana	0.90	0.99	0.98	0.96	0.97	0.98
Rotten oranges	0.77	0.93	0.92	0.88	0.95	0.92

Table 4 F-1 score values of models

Models	Base CNN	Resnet18	Resnet34	Resnet50	Resnet101	Resnet152
Fresh apples	0.90	0.98	0.97	0.95	0.96	0.96
Fresh banana	0.91	0.98	1.00	0.98	0.97	0.97
Fresh oranges	0.85	0.98	0.98	0.95	0.98	0.97
Rotten apples	0.71	0.96	0.94	0.90	0.92	0.93
Rotten banana	0.87	0.99	0.99	0.96	0.98	0.98
Rotten oranges	0.76	0.95	0.94	0.91	0.95	0.95

- Resnet18 and Resnet34 performed equivalently well for the rotten banana class.

So, based on the above analysis and also taking into account the execution time of models, we can say that our Resnet18 model performs best based on all the performance criteria considered. Resnet18 achieved the highest accuracy, took least execution time of 37 min and 49 s, and also achieved the highest precision, recall, and F1-score values for the majority of the dataset's classes.

5 Conclusions and Future Work

In our research work, the main objective was to develop a highly accurate and efficient fruits classification and grading model. For this, a base CNN model is created and, after some modifications employing transfer learning concepts, it is compared with existing pre-trained ResNet models. A proper dataset was chosen to deal with data insufficiency and data imbalance problems. The dataset contained a total of 10,200 images with each class having equal representations in the dataset. Data augmentation techniques such as center crop, random horizontal flip, random vertical flip, and rotation were applied to the dataset in order to achieve better accuracy. A detailed comparison was done among the models for the image processing tasks of classification and grading. The confusion matrix and classification report were developed. Accuracy, precision, recall, and F1-score were the performance evaluation metrics used to compare the models.

On the test dataset, the accuracy of the base CNN model was 83.8%, whereas that of the pre-trained ResNet models was over 90%. The Resnet18 and Resnet34 models achieved the maximum accuracy, 97.30%. Of all the models, Resnet18 shows the best performance in terms of all performance criteria considered such as accuracy, precision, recall, F1-score, and also execution time. The real-time image processing systems for which image processing models were created can be used to classify and grade fruits for supply chain management. This will eventually lead to a reduction in wastage and financial losses associated with it and also pave the way for industrial automation for quality inspection and management of fruits.

For future work, other pre-trained models such as AlexNet, VGG, SqueezeNet, DenseNet, etc. can be modified and compared with our models mentioned in this paper for image classification and grading tasks on the "Fruits4" dataset.

References

1. Statistics H (2018) Horticultural statistics at a glance. Horticulture Statistics Division, Department of Agriculture & Cooperation. Ministry of Agriculture & Farmer's Welfare, Government of India, p 458
2. Bogataj D, Bogataj M, Hudoklin D (2017) Reprint of "Mitigating risks of perishable products in the cyber-physical systems based on the extended MRP model." Int J Prod Econ 194:113–125

3. Katarzyna R, Pawel M (2019) A vision-based method utilizing deep convolutional neural networks for fruit variety classification in uncertainty conditions of retail sales. Appl Sci 9:3971
4. Zhang Y, Lian J, Fan M, Zheng Y (2018) Deep indicator for fine-grained classification of banana's ripening stages. EURASIP J Image Video Process 1−10
5. Yu Y, Zhang K, Yang L, Zhang D (2019) Fruit detection for strawberry harvesting robot in non-structural environment based on Mask-RCNN. Comput Electron Agric 163:104846. https://doi.org/10.1016/j.compag.2019.06.001
6. Sakib S, Ashrafi Z, Siddique MAB (2019) Implementation of fruits recognition classifier using convolutional neural network algorithm for observation of accuracies for various hidden layers. ArXiv Prepr. ArXiv190400783
7. Lu Y (2016) Food image recognition by using convolutional neural networks (cnns). ArXiv Prepr. ArXiv161200983
8. Zeng G (2017) Fruit and vegetables classification system using image saliency and convolutional neural network. In: 2017 IEEE 3rd information technology and mechatronics engineering conference (ITOEC). IEEE, pp 613–617
9. Sustika R, Subekti A, Pardede HF, Suryawati E, Mahendra O, Yuwana S (2018) Evaluation of deep convolutional neural network architectures for strawberry quality inspection. Int J Eng Technol. 7:75–80
10. Wang S-H, Chen Y (2020) Fruit category classification via an eight-layer convolutional neural network with parametric rectified linear unit and dropout technique. Multimed Tools Appl 79:15117–15133
11. Palakodati SSS, Chirra VRR, Yakobu D, Bulla S (2020) Fresh and rotten fruits classification using CNN and transfer learning. Rev Intell Artif 34:617–622

Developing Methodologies for Faster Defence Indigenisation: Opportunities for Indian SMEs

Anup Chawan and Hari Vasudevan

Abstract The Indian military base is in the process of rapid modernisation and various methodologies to achieve it have been successfully initiated by the Ministry of Defence (MoD). India is transforming its defence manufacturing capabilities by putting into action G2G collaborations positively. The 'Make in India' initiative has been globally recognised today. Indigenisation has been widely acknowledged and executed by many leading defence companies in India, like DRDO, Larsen and Toubro, MKU, Mahindra Defence Systems, Bharat Forge Limited, Tata Advanced System Limited, Ashok Leyland, etc. SMEs in India have also supported this drive up to a certain extent. However, to achieve complete indigenisation and 'Make for World', SMEs have a major role to play. In order to provide further momentum to this, this study strongly pitches two methodologies to expedite the process of indigenisation. The methodologies emphasise creating synergy between apex industry associations and other industry associations, thereby providing more opportunities for SMEs to enter the defence market. These proposed methodologies are expected to help the decision/policymakers to expand the indigenisation process at a faster rate. A supporting framework has been established further to deepen the approach towards proposed methodologies. The study is expected to further expand and encourage discussions on faster indigenisation. Finally, the study concludes with future research directions to support combat services.

Keywords Defence · Indigenisation · Indian SMEs

A. Chawan (✉)
Department of Automobile Engineering, M.H. Saboo Siddik College of Engineering, Byculla 400008, India
e-mail: chawananup@gmail.com

H. Vasudevan
Principal and Professor in Production Engineering, D. J. Sanghvi College of Engineering, Vile Parle 400056, India
e-mail: principaldjs@gmail.com

A. Sachdeva et al. (eds.), *Recent Advances in Operations Management and Optimization*, Lecture Notes in Mechanical Engineering,
https://doi.org/10.1007/978-981-99-7445-0_12

Abbreviations

ADB	Army Design Bureau
DAP	Defence Acquisition Procedure
DDP	Department of Defence Production
G2G	Government to Government
MoD	Ministry of Defence
NPD	New Product Development
OEMs	Original Equipment Manufacturers
SIDM	Society of Indian Defence Manufacturers

1 Introduction

"Market Leadership is going to come from our ability to innovate, look at the future; what could be our country's requirements, and based on that we should try & design military products" [1]. The Indian armed forces are in need of modernised indigenous weapons and equipment. MoD and SIDM have pumped fresh energy by providing a platform for interaction between the Indian tri-services (Army, Navy, Air Force), defence industry partners and academia, so as to ignite indigenisation [2]. OEMs alone cannot bring out complete self-reliance in the defence sector without the support of SMEs. SMEs are considered the most powerful weapon in the archery of a nation to bring out modernisation and indigenisation. Collective participation at both ends is essential for achieving complete autonomy for the defence sector. Several methodologies have been developed and successfully implemented by the MoD for improving indigenisation in the defence sector. There is currently a need for faster indigenisation of the defence sector as Indian armed forces are in the process of rapid modernisation. The participation from SMEs is currently less due to various challenges faced by them. One of the key challenges is the financial constraint. However, they have better technological and managerial potential. Another key challenge is the know-how required to trigger maximum participation of SMEs towards defence manufacturing.

This research study is structured into six sections. The first section throws light on the significance of indigenisation and the current scenario of defence industry in India. The second section focuses on the need for faster indigenisation as well as the objectives for the same. The third section describes the two proposed methodologies for faster defence indigenisation. The fourth section discusses the supporting framework for the proposed methodologies. The fifth section concludes with a summary of proposed methodologies for faster defence indigenisation. The sixth section finally concludes with future research directions.

1.1 Current Scenario

There have been several G2G collaborations and defence reforms with the goal of achieving indigenisation. Various indigenisation initiatives are in progress through the Make in India program. The Indian defence industries are further marching towards Make for World through interactions with different ministries of defence of other countries with the support of MoD and SIDM. Figure 1 shows the current scenario of defence industry in India.

2 Necessity of the Study

India's defence industrialisation has gone through several phases, which include the quest for self-sufficiency, self-sufficiency to self-reliance, self-reliance through co-production, self-reliance through private sector participation as well as self-reliance through Make in India initiative [3]. Figure 2 shows the different phases of India's defence industrialisation.

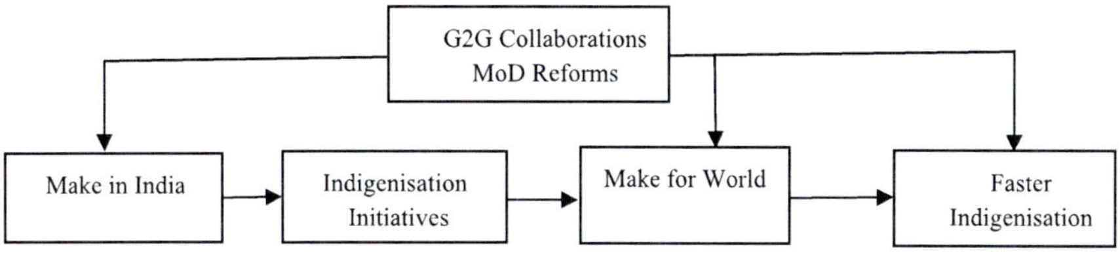

Fig. 1 Current scenario of defence industry in India

Fig. 2 Phases of India's defence industrialisation [3]

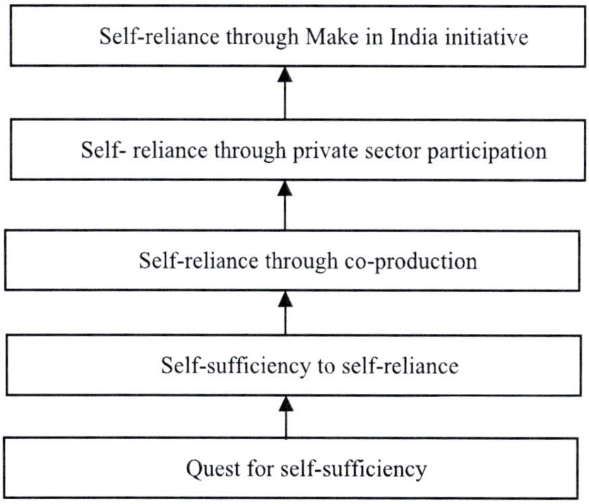

There is currently a need to indigenise faster as India aims not only to indigenise, but also to make for the world. Various schemes have been launched and 360-degree reforms have been made by the MoD [4]. 2500 items have been successfully indigenised and 351 are available on the "SRIJAN Portal" [5]. Recently 4th positive indigenisation list has been announced by the MoD with indicated timelines at Def Expo 2022 [6]. In order to speed up the indigenisation process, maximum participation of SMEs is required. Various industry associations have a greater role to play to attract SMEs towards the development of in-house defence manufacturing facilities. SMEs provide various systems, subsystems and components to cater to the needs of the Indian military. Some SMEs are also indirectly contributing by providing certain products to the defence manufacturing OEMs. However, there are SMEs, which have innovation and investment potential, but they may not be aware of new product development opportunities for the defence sector. In the coming years, they can make an entry into the defence market through proper guidance and support. In this context, the objectives of this study are as follows:

1. To develop methodologies for promoting the growth of SMEs in the Indian defence sector.
2. To develop methodologies to support faster indigenisation in the Indian defence sector.

3 Methodologies for Faster Indigenisation

In this section, the methodologies for faster indigenisation are described, as India is aiming not only towards "Make in India, but also for "Make for world" in the defence sector [7].

3.1 Methodology I

MoD has made remarkable moves in bringing out indigenisation in India. SIDM is the apex defence industry association in India [8]. It has been playing a vital role in India's journey towards indigenisation by acting as a bridge between the needs of the Indian tri-services with the defence partners in India and abroad. Several webinars have been successfully conducted by MoD, DPP and SIDM to provide B2B (Business to Business) opportunities to flourish in-house defence manufacturing facilities in India. SIDM with the support of MoD closely interacts with the Indian armed forces and organises various events to promote the growth of defence industry in India. SIDM captures the voice of the defence forces consistently [8]. Figure 3 shows the first methodology for faster indigenisation. The A, B, C letters indicate different types of industry associations in the Fig. 3.

Various problem definition statements have been provided by the ADB [9]. Please refer to Appendix A for the list of problem definition statements. They have been

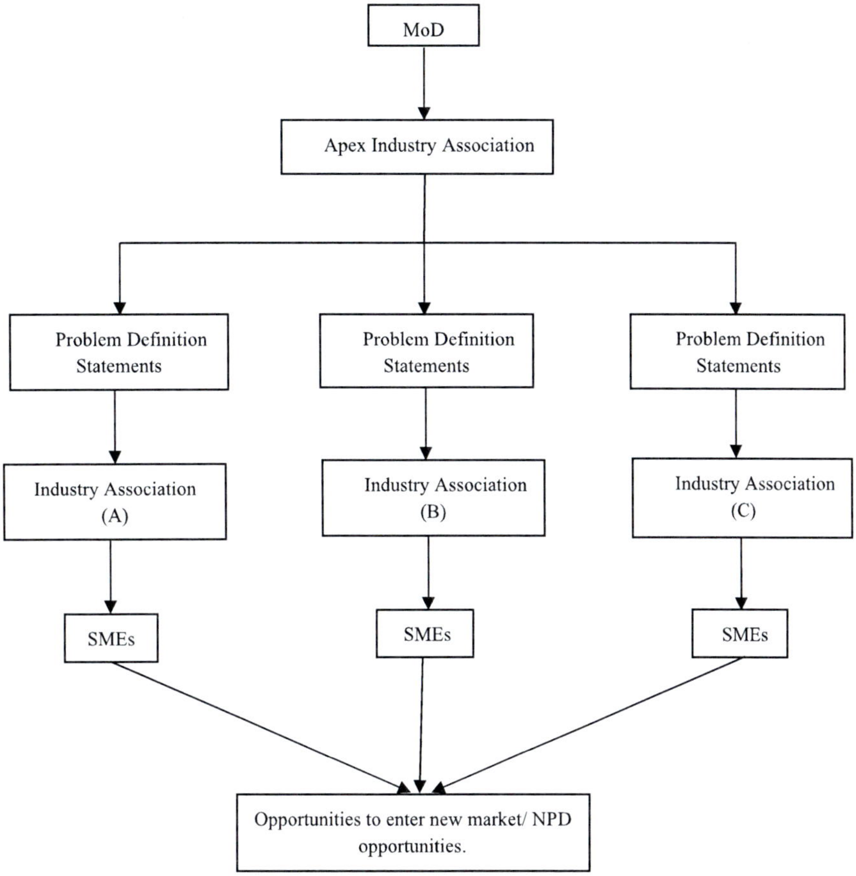

Fig. 3 Methodology I for faster indigenisation (Authors' perspective)

specially formulated to provide indigenous solutions in terms of research & development, design & development, prototype, limited series production and mass production [10]. These problem definition statements are available in the public domain. Any Indian SME, research institute, academia, etc. can provide indigenous solutions [9]. Micro Small and Medium Scale Enterprises (MSMEs) will be offered different projects costing less than Rs 100 crore under the DAP 2020 provisions [10]. 70 solutions out of 130 problem statements have been adopted under various projects [11].

There are various in general and sector-specific industry associations in India. They carry different types of activities to promote the growth of the SMEs. The industry associations are very well aware of the strengths and weaknesses of SMEs as they are active participants in various talk shows and exhibitions organised by the industry partners. The remaining undeveloped or unchallenged problem statements can be showcased to the SMEs by creating a synergy between the apex defence industry association with other industry associations in India. The synergy between them is expected to give healthy opportunities for SMEs to make an entry into the defence sector/ NPD opportunities.

3.2 Methodology II

The MoD had announced a total of three positive indigenisation lists consisting of 310 items earlier [6]. Recently the MoD released the fourth indigenisation list in October 2022 [12]. Refer to Appendix B for all the lists of positive indigenisation items. Recently the SMEs definition has been revised based on the turnover [13].

This fundamental change will allow many firms to come under the SMEs category, which makes it easy to avail the benefits under the provisions of DAP 2020. This shows the proactive approach of MoD. The idea behind this proposed methodology is to maximise SMEs participation and take advantage of the provisions of DAP 2020.

However, to achieve this faster, a synergy among the apex defence industry association and other industry associations is essential. In order for SMEs to indigenise, they must be made aware of different product/technology development opportunities. Synergy among the various industry associations is going to play a key role in the process of indigenisation by creating awareness, guidance and support for SMEs. Figure 4. shows the second methodology for faster indigenisation. The A, B, C letters indicate different types of industry associations in Fig. 4.

4 Supporting Framework for Proposed Methodologies

Figure 5 shows the framework for supporting the proposed methodologies for faster defence indigenisation. The study recommends to establish synergy among select industry associations belonging to a few of the champion sectors, such as electronics i.e. Electronics System Design & Manufacturing (ESDM), Automotive & Auto Components, Chemical & Petrochemicals and Textile & Apparels with the apex industry association. The aim of establishing the synergy between the select industry association and apex industry association is to identify different project opportunities by exploring various indigenisation lists and problem statements listed in Army Design Bureau (ADB). As discussed earlier in the defence acquisition procedure 2020, the projects costing less than 100 crores will be procured from SMEs. Select manufacturing SMEs should be offered such projects based on the selection criteria as listed in Fig. 5. It includes innovation awards won by them, other awards and recognitions from OEMs, proven technological innovativeness and manufacturing performance in at least of last 3 to 5 years. Also, the SMEs adopting/developing technologies relevant to defence sector should be encouraged to take over indigenisation products further. Existing export performance, domestic sales and patents are to be

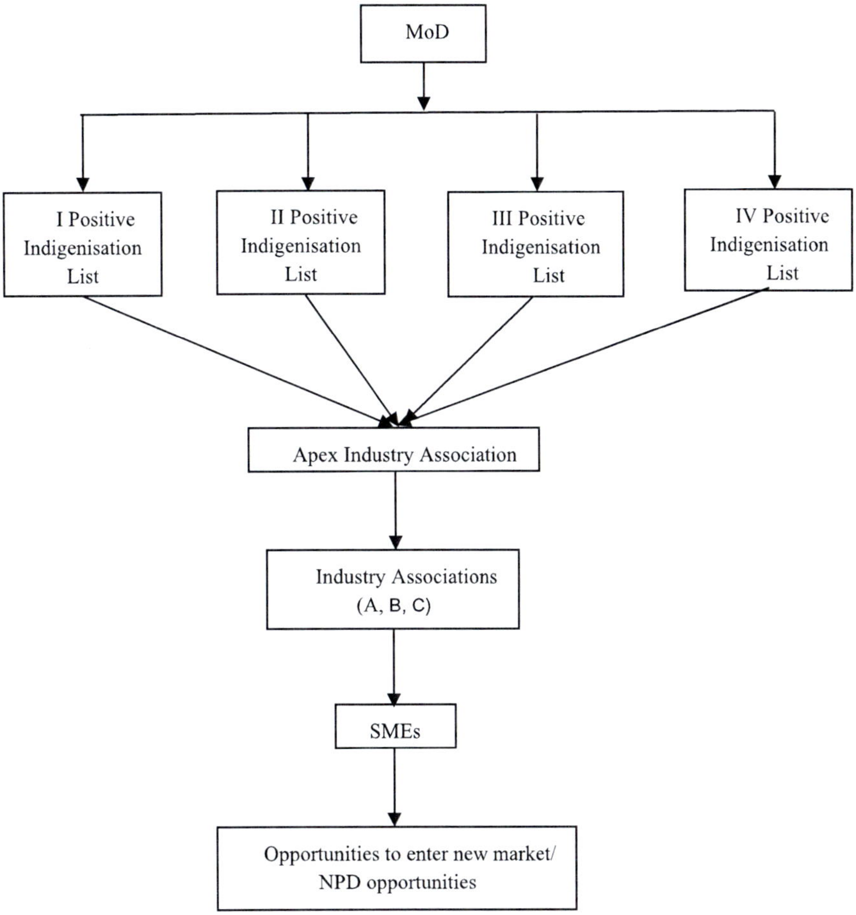

Fig. 4 Methodology II for faster indigenisation (Authors' perspective)

taken into consideration for project allocations. As discussed, this all brings various opportunities to enter new markets for the SMEs in defence sector.

5 Discussion

India has already started strengthening its bilateral relations with other countries to promote indigenisation. Many strategic collaborations are in progress. Recently, the Czech Republic has initiated the process of forming a joint venture with India during their recent visits to India at Def Expo 2022 [14]. Many B2B opportunities have been successfully triggered by the MoD of India with the MoD of various other countries. Not only OEMs, but also some manufacturing firms registered under the Ministry

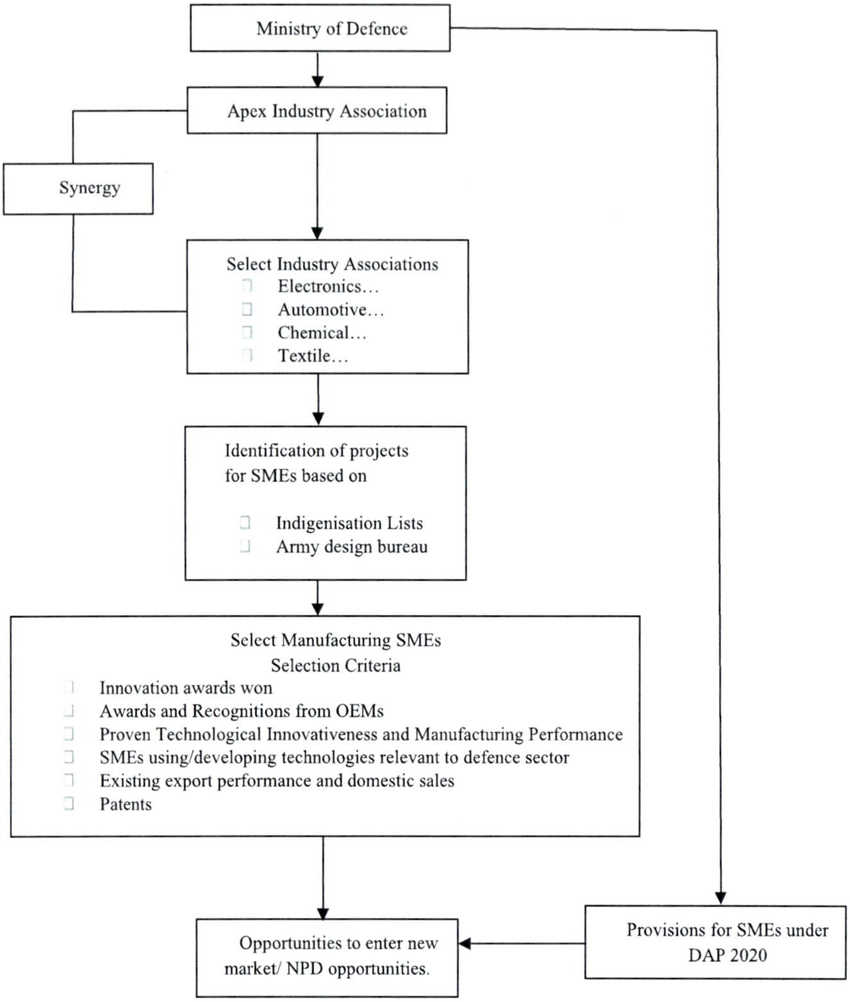

Fig. 5 Framework for supporting proposed methodologies (Authors Perspective)

of MSMEs have showcased their product portfolio in export webinars organised by MoD and SIDM [15]. The 'Make in India' initiative has been widely recognised by the defence ministries of other countries too. Since this initiative was launched, other countries have started showing their interest to invest in India as well as to collaborate for mutual benefits, especially in the defence sector. The embassies of several countries have supported this initiative by encouraging bilateral relations to promote defence sector. Defence corridors have been formed in India in two states. i.e., Uttar Pradesh and Tamil Nadu to promote the 'Make in India' initiative [16]. As discussed earlier, developing synergy between the apex defence industry association and other industry associations is of prime importance. Many states in India are known for their core competencies.

For aerospace industries, Tamil Nadu is well known. In order to promote the growth of aerospace and defence sector in Tamil Nadu, emphasis has been given to bringing all the industry associations together by creating Tamil Nadu specific association [17]. As stated earlier, there are many industry associations in India, such as the Automotive Component Manufacturers Associations of India (ACMA) and the Tool & Gauge Manufacturers Association of India (TAGMA). Many industry associations have emphasized on indigenisation and developing synergy between apex industry associations with other industry associations is the need of the hour for cultivating faster Defence indigenisation. Brainstorming is certainly required to channel as to how the synergy between apex industry association and other industry associations can be utilised effectively. Many senior officials from defence OEMs and MoD have stated that through collective efforts India can indigenise in defence sector. In both the methodologies described in this research paper, the central theme is as to how the synergy between the apex defence industry association and other industry associations would help us to accomplish the goal of indigenisation.

6 Conclusion and Future Scope

This research work has emphasised on the need for faster indegenisation and recent happenings in the Indian defence sector and has made few contributions by proposing methodologies to foster indigenisation. In this context, two conceptual methodologies have been established, focusing on problem definition statements and lists of positive indigenisation items. The importance of synergy among industry associations has been discussed and the study has addressed the need for promoting SMEs in the defence sector. A supporting framework was established to further deepen the approach towards the proposed methodologies. In order to gain a clear understanding and execution of defence indigenisation, it is essential to associate with apex industry associations and other associations, so that SMEs can expand their product mix by entering the defence market through proper guidance and support. 'Synergise to indigenise' is the key to achieve indigenisation in the defence sector.

Future research could focus on, as to how the synergy among industry associations would help SMEs explore opportunities in terms of dual use of technology applications and at the same time develop products at competitive costs. For example, technology, such as the Internet of Things (IoT) has applications in many sectors, such as automotive, healthcare, defence. The technology is already established for military applications, called as military Internet of Things (MIoT). There are other technologies, like blockchain technology and quantum computing, etc., having applications in the defence sector and other sectors too and there are lot many opportunities for SMEs to explore. It is expected that SMEs with the support of industry associations, OEMs and academia come together to ensure India's journey towards defence indigenisation in a timely and effective manner.

Acknowledgements The authors of this research paper would sincerely like to thank the Ministry of Defence (MoD) India and the Society of Indian Defence Manufacturers (SIDM) for allowing us to be part of various international bilateral programs, focusing on indigenisation as well as other international and national programs through online platforms. Special thanks to the Director General of SIDM for his kind consideration and invitation given to us to share various inputs.

Appendix A

The problem definition statements are available on the link given below:

https://indianarmy.nic.in/writereaddata/adbdocuments/Compendium_of_Problem_Definition_Statements_2020.pdf.

Appendix B

The positive indigenisation items are available on the link given below:

Positive Indigenisation List I.
https://srijandefence.gov.in/DPSU%20Indigenisation%20List.pdf

Positive Indigenisation List II.
https://www.mod.gov.in/sites/default/files/OM%20LI04621.pdf

Positive Indigenisation List III.
https://srijandefence.gov.in/DPSUs3rdPIL.pdf

Positive Indigenisation List IV.
https://mod.gov.in/sites/default/files/4th-PIL_comp.pdf.

References

1. Joint Secretary, Government of India (GoI), Department of Defence Production. https://nenp.facebook.com/SIDMIND/photos/a.139715256722497/1045947209432626/. Accessed 17 Feb 2023
2. Chief of Army Staff, Department of Defence Production.: Synergising Defence Industry competence in Research & Design with National Security Need. https://www.youtube.com/watch?v=UwPgvBpsxJc. Accessed 01 Dec 2022
3. Behra LK (2016) Indian defence industry: an agenda for making in India. Pentagon Press and Institute for Defence Studies and Analyses, New Delhi
4. Ministry of Defence, Department of Defence Production, Government of India: Propelling Private Sector Participation. https://static.pib.gov.in/WriteReadData/specificdocs/documents/2021/sep/defenceenglish.pdf. Accessed 01 Jun 2022
5. Ministry of Defence, Department of Defence Production, Government of India. https://srijandefence.gov.in/DPSU%20Indigenisation%20List.pdf. Accessed 05 Dec 2022

6. Ministry of Defence, Department of Defence Production, Government of India. https://www.mod.gov.in/sites/default/files/PM-announces-fourth-positive-indigenisation-list.pdf. Accessed 07 Dec 2022
7. Defence Minister, Government of India. https://economictimes.indiatimes.com/news/defence/make-defence-products-for-world-rajnath-singh-tells-industry/articleshow/94490395.cms. Accessed 07 Dec 2022
8. Society of Indian defence manufacturers (SIDM). https://www.sidm.in/. Accessed 02 Dec 2022
9. Directorate General of Perspective Planning, Army Design Bureau, Compendium of Problem Definition Statements 2020. https://indianarmy.nic.in/writereaddata/adbdocuments/Compendium_of_Problem_Definition_Statements_2020.pdf. Accessed 15 Oct 2022
10. Make in India. https://indianarmy.nic.in/Site/FormTemplete/frmTempSimple.aspx?MnId=pkhmDdUbxYWYX104D/H7FkGV5iQddTJBLaOd/gXQ+pE=&ParentID=AIt1PZMbqVZsquMdG4vZSw==&flag=8CKP966uzg96kLov0aWdfQ. Accessed 15 Oct 2022
11. Chief of Defence Staff (2020) Directorate General of Perspective Planning, Army Design Bureau, Compendium of Problem Definition Statements 2020. https://indianarmy.nic.in/writereaddata/adbdocuments/Compendium_of_Problem_Definition_Statements_2020.pdf. Accessed 15 Oct 2022
12. President, SIDM, Department of Defence Production, Synergising Defence Industry competence in Research & Design with National Security Need. https://www.youtube.com/watch?v=UwPgvBpsxJc. Accessed 15 Dec 2022
13. SME chamber of India. https://www.smechamberofindia.com/about-msme-in-india.php. Accessed 15 Dec 2021
14. Czech Ambassador to India (2022) Webinar on India-Czech republic defence industry cooperation, Ministry of defence, Department of defence production, Government of India
15. Webinar on India-Czech Republic Defence Industry Cooperation, Ministry of Defence, Department of Defence Production, Government of India (2022)
16. Defence Industrial Corridors in India. https://www.makeinindia.com/defence-industrial-corridors-india. Accessed 15 Dec 2021
17. Senior Manager Projects (TIDCO).: Tamil Nadu the emerging hub for aerospace & defence industries, ministry of defence, Department of Defence Production, Government of India. https://www.youtube.com/watch?v=7rH-w_V1sHw. Accessed 20 Jan 2023

Developing a Predictive Model to Identify and Analyse Vehicle Attributes in the Indian Automotive Supply Chain Using Big Data Analytics and MCDM Techniques

Ruben Kuruvilla Thomas, Rinu Sathyan, Sandeep Sunil, and A. S. Abin

Abstract Automotive supply chain has undergone a shift in customer preferences towards vehicle features. The COVID pandemic increased demand for personal mobility thus affecting the Indian economy and changing customer attitudes towards cars. Also, the critical part shortages caused repercussions throughout global supply chains. Due to this, it is now necessary to analyse vehicle attributes. For each new demand, the supply chain's response should be fresh and distinct. Supply chain companies are working to improve efficiency and cost-effectiveness. BDA can aid SCM by analysing customer preferences. This study aims to enhance responsiveness by using big data and MCDM techniques for analysing vehicle attributes. The findings of this research will help supply chain professionals form strategies for improved responsiveness.

Nomenclature

SCM	Supply Chain Management
BDA	Big Data Analytics
ISM	Interpretive Structural Modelling
MCDM	Multi-Criteria Decision-Making
SSIM	Structural Self-Interaction Matrix

R. K. Thomas (✉) · R. Sathyan · S. Sunil · A. S. Abin
Department of Mechanical Engineering, Sree Chitra Thirunal College of Engineering, Trivandrum 695018, India
e-mail: ruben.thomas456@gmail.com

A. Sachdeva et al. (eds.), *Recent Advances in Operations Management and Optimization*, Lecture Notes in Mechanical Engineering,
https://doi.org/10.1007/978-981-99-7445-0_13

1 Introduction

A group of people connected by the exchange of goods, information, and money with the aim of bringing down the cost of the entire system is termed the "supply chain" [1]. The supply structure can vary depending on the number and location of intermediary facilities in the supply chain [2]. Supply chain responsiveness refers to the ability to adapt quickly to changing market needs, which is crucial for addressing variables that affect consumer satisfaction [3]. In the automotive industry, efficiency has traditionally taken precedence over responsiveness, but this approach is changing due to the introduction of vehicles with enhanced features and attributes. By identifying attributes, manufacturers can improve their understanding of how product features affect consumer interest and adjust their production accordingly, leading to a more responsive supply chain [4]. The concept of responsiveness challenges established procedures in product creation, production, and logistical planning. Big data analytics (BDA) is an emerging tool for supply chain management and logistics, with applications in demand planning and forecasting based on historical data.

We employed the multi-criteria decision-making (MCDM) technique in this study to replicate the factors that influence the responsiveness of the automotive supply chain. The vehicle attributes were identified through sentiment analysis and expert opinion. The causal relationship between the vehicle attributes was developed using the interpretive structural modelling (ISM) approach and the attributes were categorized using MICMAC analysis. The managerial implications presented in this paper will offer valuable insights to supply chain practitioners, enabling them to develop effective strategies for enhancing responsiveness.

2 Methodology

The various attributes of the vehicle should be determined, and a qualitative evaluation is done to identify the interrelationship between them [5]. In this study, we will apply a combined big data analytics methodology to analyze vehicle attributes using the MCDM technique. The core base that will provide the vehicle attributes are the online review platforms such as Zigwheels and Overdrive, which will go through web scraping to provide the entire reviews of the vehicle by the users. Python code will then be used to take the word frequency from the extracted reviews which will lead to the vehicle attributes.

The vehicle attributes will be chosen by a group of automotive industry experts once the word frequency has been collected. These experts will also explain how the attributes relate to one another in order to use the ISM technique. An ISM matrix will then be developed to understand how the attributes relate to one another. These parameters were examined using MICMAC Analysis based on their driving and dependency factors.

Table 1 Sample of reviews of the vehicle and its rating

Serial number	Review	Rating
1	Super for middle-class people and safety security as well best price	4
2	This is the best car in this budget and 'X' always stands for safety we all know	5
3	My first car safety car overall nice car	3.7

2.1 Web Scraping Using Parse Hub

A web scraping tool called Parse Hub was used to gather information from a variety of automotive-related websites, including Zigwheels, Overdrive, CarWale, and many others. Parse Hub is used to extract vehicle reviews. The retrieved data was then exported to an Excel sheet, from where they will be later exported for further investigation. A sample of the review collected is given in Table 1.

2.2 Word Frequency Analysis Using Python Program

A sentiment analysis was conducted on the extracted reviews using a Python code to categorize them as either 'positive,' 'negative,' or 'neutral.' Subsequently, the focus was narrowed down to collecting only the positive reviews to ascertain the most influential attributes of the vehicles under consideration. The top fifty words that were connected to the car were selected from this output according to the opinion of industry experts. After evaluating these fifty words, the most influential vehicle attributes were chosen for conducting further studies. This selection of nine attributes was done by experts from the field. The vehicle attributes selected for further study are depicted in Table 2.

Table 2 Vehicle attributes and code to represent

Code	Vehicle attributes
A1	Mileage
A2	Safety
A3	Engine performance
A4	Comfort
A5	Maintenance
A6	Interior
A7	Security
A8	Power
A9	Modern facilities and specifications

2.3 *Interpretative Structural Modelling Technique*

ISM is a method that establishes the structural relationship between selected factors and establishes a structural division of various levels into which each of these components falls. It is a philosophical form of modelling [6].

1. *Step 1*: Identifying vehicle attributes of Supply Chain Responsiveness.
2. *Step 2*: Determine the relationship between the cited variables based on expert judgement.
3. *Step 3*: After developing the ISM Matrix, an SSIM Matrix is formed by filling the matrix with either of the 4 symbols; V, A, X or O.

    - V: For attributes 'a' and 'b', factor 'b' can be found using 'a', but vice-versa is not possible.
    - A: For the same attributes, 'a' can be found using 'b', but vice-versa is not possible.
    - X: Both the elements could be found out using each other.
    - O: Both the elements cannot be calculated with the help of each other.

4. *Step 4*: Forming initial and final reachability matrix

The initial reachability matrix is created by substituting symbols with 1s and 0s. After determining if the third attribute is related to the second attribute, which is related to the first attribute, and if so, the third variable will be related to the first variable. By assessing the relationships between attributes, the final reachability matrix is obtained from the initial one. The variables are subsequently classified into levels according to the reachability matrix's final outcomes [7].

5. *Step 5*: Level partitioning

The process involves determining both the Reachability set and the Antecedent set for each attribute, followed by recording the intersection of these sets for each attribute. Initially, at level 1, the attribute demonstrates equality between its reachability set and intersection. Once there are no more such attributes forming level 1, the attributes that formed the initial level are removed and the intersection is done for the remaining variables. In this manner, the level partition is done with the entire variables.

2.4 *MICMAC Analysis*

The driving power and dependency factor of each attribute can be derived from the final reachability matrix. These values are used for MICMAC Analysis, which involves plotting the attributes on a graph. Attributes are categorized into four quadrants based on their driving power and dependence factor.

- Autonomous Enablers possess limited driving power and exhibit a reliance on external factors.

- Dependence exhibits limited driving force but demonstrates significant dependence.
- Linkage enablers possess significant driving forces and exhibit interdependence.
- Independent enablers possess robust motivating forces but exhibit limited interdependence.

3 Illustration of the Proposed Methodology in a Case Vehicle of an Automotive Supply Chain in India

In India, a case study was carried out on a compact sports utility vehicle (SUV) to showcase the effective integration of BDA and ISM methodologies within the automotive industry. There are nine automotive manufacturing companies in this sector. The study used 'Model X', which was launched in October 2021, and by November 2022, approximately 46,000 cars had been sold, priced between Rs. 6 lakhs and Rs. 9.6 lakhs.

4 Result and Discussions

4.1 Developing the ISM Model and its Evaluation

The ISM technique aims to develop drivers' hierarchical level, using an SSIM (Table 3) learned from industry experts to represent driver interactions. The matrix was completed with V, A, X, and O based on the relationship of one factor to another. The resulting reachability matrix (Table 4) provides information on driving power and dependence for each vehicle attribute [6] (Table 5).

Table 3 Structural self-interaction matrix (SSIM)

Factors	Code	A1	A2	A3	A4	A5	A6	A7	A8	A9
Mileage	A1	X	O	A	O	A	O	O	O	A
Safety	A2		X	O	O	X	O	A	O	A
Engine performance	A3			X	O	A	O	O	X	A
Comfort	A4				X	A	X	O	O	A
Maintenance	A5					X	V	A	V	O
Interior	A6						X	O	O	A
Security	A7							X	O	A
Power	A8								X	A
Modern facilities and specifications	A9									X

Table 4 Final reachability matrix

Code	A1	A2	A3	A4	A5	A6	A7	A8	A9	Driving Power
A1	1	0	0	0	0	0	0	0	0	1
A2	1*	1	1*	1*	1	1*	0	1*	0	7
A3	1	0	1	0	0	0	0	1	0	3
A4	0	0	0	1	0	1	0	0	0	2
A5	1	1	1	1	1	1	0	1	0	7
A6	0	0	0	1	0	1	0	0	0	2
A7	1*	1	1*	1*	1	1*	1	1*	0	8
A8	1*	0	1	0	0	0	0	1	0	3
A9	1	1	1	1	1*	1	1	1	1	9
Dependence	7	4	6	6	4	6	2	6	1	

Table 5 Level partition

Attribute	Reachability set	Antecedent set	Intersection set	Level
A1	A1	A1, A2, A3, A5, A7, A8, A9	A1	I
A4	A4, A6	A2, A4, A5, A6, A7, A9	A4, A6	I
A6	A4, A6	A2, A4, A5, A6, A7, A9	A4, A6	I
A3	A3, A8	A2, A3, A5, A7, A8, A9	A3, A8	II
A8	A3, A8	A2, A3, A5, A7, A8, A9	A3, A8	II
A2	A2, A5	A2, A5, A7, A9	A2, A5	III
A5	A2, A5	A2, A5, A7, A9	A2, A5	III
A7	A7	A7, A9	A7	IV
A9	A9	A9	A9	V

Based on the ISM approach, vehicle attributes were classified into five levels (Fig. 1). Mileage (A1), Comfort (A4), and Interior (A6) factors form the first level. This indicates that these drivers possess strong driving abilities. Engine Performance (A3) and Power (A8) form the second level, followed by Safety (A2) and Maintenance (A5) forming the third level. Security (A7) is on the fourth level and Modern facilities and Specifications (A9) is on the final level. Using this model, automobile manufacturers and supply chain professionals can improve responsiveness with effective strategies [6].

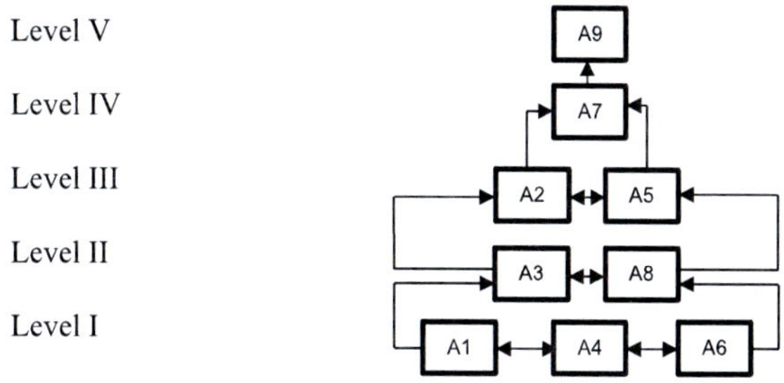

Fig. 1 Level partition

4.2 MICMAC Analysis

It is a study that was conducted to support the ISM method. It divides many factors into four quadrants with distinct significance [8]. Figure 2 shows the plot of each attribute on the graph based on the driving power and dependence factor.

Cluster IV, which comprises Maintenance (A5), Security (A7), Safety (A2), and Modern facilities and Specifications (A9), is a vital component of management improvement, with a strong driving force and weaker dependence. In contrast, Cluster I does not include any features and is categorized by weak driving power and weaker dependence. Cluster II, on the other hand, encompasses Mileage (A1), Engine Performance (A3), Comfort (A4), Interior (A6) and Power (A8), all of which are critical to both the supply chain and the vehicle.

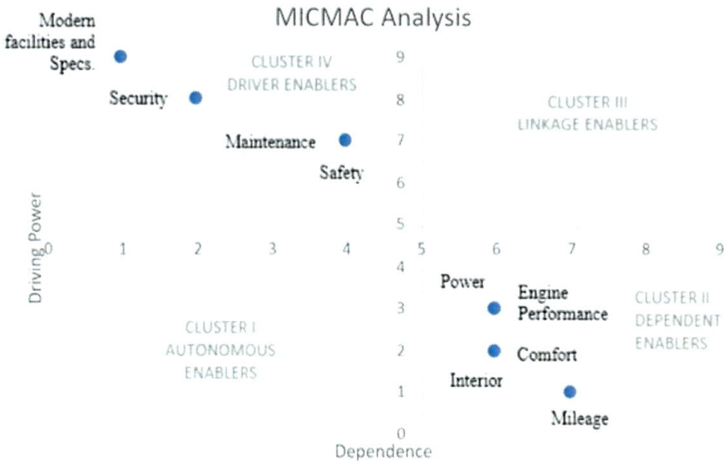

Fig. 2 MICMAC analysis

5 Managerial Implications

The pandemic has led to a rise in private vehicle ownership, prompting the automobile industry to shift its focus to passenger vehicles. Various factors can impact personal vehicles, with each component having different effects. Successful vehicle sales depend on choosing the right attributes based on customer needs and price range. The ISM technique, used to determine buyer priorities and interdependence of factors, highlights the importance of an adaptable and responsive supply chain. Hierarchical division and level partition provide management with an understanding of how factors are interconnected. The study's MICMAC Analysis clusters factors and emphasizes the need to consider what motivates customers to purchase goods. The supply chain's ability to respond to these factors without negative business impact determines its success. The supply chain's responsiveness to changes in automobile features, based on customer requirements, is critical for future trends. The study's analysis can help management develop strategies to improve supply chain responsiveness.

6 Conclusions

The study aims to identify key vehicle attributes that affect purchase intent and improve automotive supply chain responsiveness. Online reviews are analysed using sentiment analysis and the ISM-MICMAC technique with input from industry experts. However, there are limitations, such as the inclusion of fake reviews and the limited scope of the study. Future research will explore additional factors that influence the automobile supply chain. Advanced decision-making methodologies can be used to identify the relationships and relevance of the drivers.

References

1. Simchi-Levi D, Kaminsky P, Simchi-Levi E (2008) Designing and managing the supply chain: concepts, strategies and case studies, 4th edn. Tata McGraw-Hill, pp 1–3
2. Asamoah D, Nuertey D, Agyei-Owusu B, Akyeh J (2013) The effect of supply chain responsiveness on customer development. Int J Logist Manag 32(4):1190–1213
3. Gunasekaran A, Ngai EWT (2005) Build-to-order supply chain management: a literature review and framework for development. J Oper Manag 23(5):423–451
4. George J, Pillai MV (2018) A study of factors affecting supply chain performance. J Phys: Conf Ser 1355:1–2
5. Sathyan R, Parthiban P, Dhanalakshmi R, Minz A (2021) A combined big data analytics and Fuzzy DEMATEL technique to improve the responsiveness of automotive supply chains. J Ambient Intell Hum Comput 7949–7963
6. Sathyan R, Palanisamy P, Suresh G, Navin M (2022) Modelling the drivers of responsiveness of automotive supply chain using an integrated fuzzy DEMATEL-ISM approach. J Glob Oper Strat Sourc 2398–5364

7. Pathak P, Singh MP (2019) Barriers analysis for sustainable manufacturing implementation in Indian manufacturing industries using interpretive structural modelling
8. Ahmad M, Tang XW, Qiu JN, Ahmad F (2019) Interpretive structural modeling and MICMAC analysis for identifying and benchmarking significant factors of seismic soil liquefaction

A Comprehensive Review of Lean Warehousing Design Methodologies

Dominic Frappier, Hartaj Singh, Kapil Singh, and Sachit Vardhan

Abstract Lean warehousing is a formulation that extends resourcefulness consumption without loss of quality and productiveness. Lean warehouses strive to do more with a minor, thereby streamlining the working operations. Lean warehousing statement that allows businesses to deliver the goods. Lean warehousing is a commercial activity in which the use of time, functional area, and resources in a warehouse are made more efficient through automation and proactive planning, enhancing customer satisfaction. According to the lean intellect, the performance and efficiency of warehouse operations depend on the layout planning, material handling methods, and mode of transportation. In this paper, the contemporary literature on the entire methodology of the lean warehouse structure and related tools is explored, used for reducing inefficiencies and improving work productivity from the significance of customer satisfaction. The proposed gaps would afford an emerging road map for research in existing and newfound unexplored directions in lean warehousing operations.

1 Introduction

Nowadays, factors such as economic processes, assertion, rapid market changes, short goods life cycles, mass production, and decrement in waste and errors reordering the outcome of logistics on production are far more comprehensive than previously used in design processes [1, 2]. Such an analyzable hypothesis becomes a significant involvement in the design, formulation, and control of warehousing plans of actions as modern research content. Warehouse material handling facilities are used to receive, inventory, order-picking, group, reorder, and shipping of products [3]. Standardized conception allows workers to assign duties in a similar order every time and set a

D. Frappier
Department of Distribution Groupe Marcelle, Montreal, Canada

H. Singh (✉) · K. Singh · S. Vardhan
Department of Mechanical Engineering, CT University, Ludhiana, Punjab 142024, India
e-mail: hartajsinghae@gmail.com

A. Sachdeva et al. (eds.), *Recent Advances in Operations Management and Optimization*, Lecture Notes in Mechanical Engineering,
https://doi.org/10.1007/978-981-99-7445-0_14

standard. The formed process combines and applies different resources significantly, such as technology, time, tools on display boards, and unprocessed materials [4]. Process standardization is a lean physical exertion intended as a part of waste control by implementing lean manufacturing to accomplish customer satisfaction [5]. Lean warehousing can be routed as a major differential for adopting tools that control the movement of storage that enhance customer services and the level of customer satisfaction [6]. Many researchers have explored the process standardization derived using a fundamental approach of lean warehousing to make logistics collaboration between warehousing and different functional areas. Process standardization is a helpful lean approach for controlling errors in warehousing operations as it provides a set of classical instructions that can assist in reducing inventories in warehousing operations [7]. The applications of lean in warehouses present ample possibilities for upcoming research that developing new tools and models in the future. Lean warehousing operations reduce time and financial loss, by removing wastage to a state of achievable and optimizing domains that add value to the consumer. Lean warehousing assures bringing quality to consumers due to inflated efficiency and productivity vantage from lean activities and values. Although there is no appropriate definition of lean warehousing, based on some thinking and activities, it works and that, when implemented, power the components of a warehouse [8]. These thinking-based activities address all the structural, operational, and human characteristics accompanying lean warehousing. The design proposals are also acknowledged in the literature considering lean warehousing [9]. The effect of the seven wastes such as shortcomings, unnecessary inventory, overproduction, redundant motion, waiting, transporting, and incompatible processing) on warehouse functional performance, can be framed as a design to guide practices to minimize wastage in warehouses [10]. Therefore, to meet the consumer and manufacturer views, warehouse processes must be optimized, removing skillfulness and making them trustworthy in the state of cost [11, 12]. Therefore, the standing of warehouse layout and continuous enhancement arise in industrial organization and standards. In this context, the authors pretend a conceptualization addressing such a future need for applying lean approach-based operations in warehouses. Therefore, insistent on the success of a business organization warehouses are to be designed with the psychological feature of being cost-effective.

2 Literature Review and Design Methodologies

(Gunasekaran, 1999) investigated a new framework to enhance the potential of warehousing processes and accomplish cost reductions with high-quality finished goods. In this context, the authors use lean conceptions such as pulled flow, JIT approach, waste a minimal, minimum lot, and continuous enhancement, aiming to reduce storage in operation, operation times, and to guarantee greater operational changeableness [13]. (Alsmadi et. al. 2012) investigated the hypothesis postulated that organizations could deliver goods savings of up to fifty percent via the standardization application of the lean concept in warehousing [14]. (Abushaikha et al. 2018) focusing on lean warehousing, the authors researched to determine the issue of warehouse waste diminishing exercises on the company's operation. Then, validate that operational and individual performance is enhanced with the lean approach [15]. (Schonberger 1986) studied and believed to be cleanly centralized on eliminating seven important wastes and on regarding customers, workers, and vendors [16]. (Sanchez and M. Perez 2004) evaluated screening contrasting dimensions of lean by considering indicators of the lean approach [17]. (Malmbrandt and Ahlstrom 2013) investigated the developed and logical in an iterative procedure to evaluate lean work adoption [18]. (Abdi et al. 2006) studied the use of lean in four dynamic components such as learn, evaluate, analyze, and navigate [19]. (Apte and Goh 2004) have observed the level of lean services acceptance to reduce stock and lead time [20]. (Malladi et al. 2011) studied eight wastage in services such as inventory, imperfectness, motion, waiting, transportation, over-processing, over-production, and human effort [21]. In the warehouse design process, in production units, different tools, and techniques were commonly exploited for each phase. These tools and techniques were then related to those cited in the literature that has already been explored by many authors. Furthermore, database investigations were channeled on these tools and techniques so that a gap in the literature applicable to the content of lean warehouse design was covered. Figure 1 illustrates that the warehouse performance-based methodology can maximize the results from the starting point to the end user with the aid of corrective action tools for immediate decisions made without wasting time and by using people-value delivery and real-time decisions.

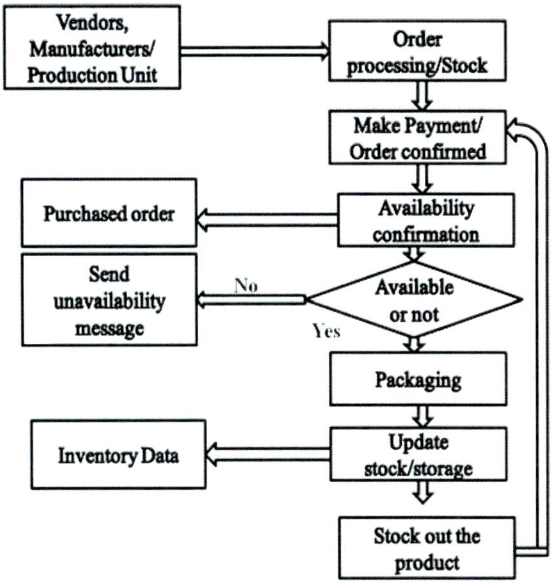

Fig. 1 Layout of warehouse performance flow sketch

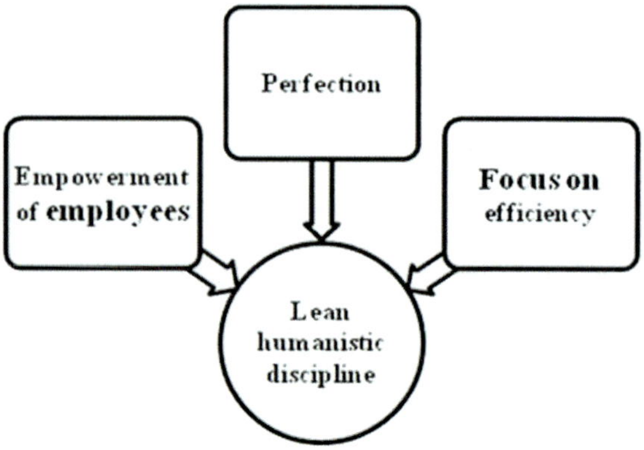

Fig. 2 Lean humanistic discipline

2.1 Warehousing Processes and Lean

(Dhaouadi et al., 2022) studied warehouses and processes that can be found in a typical warehouse via the lean thinking concept. Several different equipment and design features can improve the work performance of the lean warehousing approach [22]. (Rebelo et al. 2021) studied that lean warehousing effectively maintains the problem of stock-out in the inventory and updated time-to-time picking and receiving the movement of products by registering either manually or information database. Several strategies are used by many practitioners and create error-free storage in warehouses [23]. (Namazi 2009) explored the famous methods for mass production products of storage picking position commonly used such as the no method, the ABC method, and the family group method. In the ABC method, a warehouse is classified into three zones: A zone refers to fast-moving goods, B zone to the medium-moving and C zone for slow-moving goods. The family grouping method comprises groups having the same features such as the same customer, the same type in a mass production state, and the physical appearance of goods, and such approaches are closely relevant to lean thinking, which is broadly used in warehouses [24]. (Shafeek et al. 2018) studied the sorting time-saving approach, and the warehouses have multiple orders of picking strategies employed in this context. After that, picking the goods the packing process is to be scheduled in the warehouse for the sake of some areas or stations specially dedicated for packing purposes. The process of shipping in the warehouses themselves allows the goods they must organize distribution themselves [25]. (Raghuram and Arjunan 2022) explored three components, an increase in warehouse quality, productivity, staff, and customer satisfaction can improve the overall performance of the design framework for a lean warehouse [26].

2.2 Lean Warehouse Tools

The various lean thinking-based tools can resolve several issues by using different approaches to warehouse performance. (Metha and Dave 2020) standardization refers to the 5S technique [27], (Singh et al. 2009) continuous improvement refers to Kaizen [28], (Jaffar et al. 2015) reducing waste refers to TIMWOOD assessment of seven wastages [29], (Kumar et al. 2016) error proofing refers to POKA-YOKE [30], (Carvalho et al. 2019) described the enhancing activities flow known by VSM (value stream mapping) [31], (Amrina et al. 2019) analyzing and solving root causes refers to fishbone [32], and (Patel et al. 2021) investigates the Pareto rule to enhance inventory allocation [33]. (Wolniak et al. 2018) Bottleneck analysis is used to detect the exact point causing the process performance [34]. (Grzelczak et al. 2016) Muda removal eliminates the wastefulness of anything in the operation or supply chain that does not increase value from the consumer's perspective [35]. (Vargas et al. 2018) PDCA stands for Plan-Do-Check-Act and it comprises four steps a new thinking tool-based systematic approach for enhancing new or improved design or processes, goods or services, and to finding problems [36]. (Dalgobind and Anjani 2008) Root cause analysis (RCA) form of methodology-based practice that is used to determine the underlying issues. The process of finding the root causes of problems and identifying, recommending, and implementing the best solutions [37]. (Bilalis et al. 2002) Visual factory visualization works on the clean living of lean manufacturing. A visual factory comprises displays, graphics, charts, icons, and controls that are used to enhance the processes by applying the system of communication tools to share information on the company end [38]. (Rahman et al. 2013) Kanban thinking is used to improve the performance of the company by visualizing both prospects it includes information on the flow of production at every stage and the position of each line of work performed by employees [39]. (Jiju et al. 2017) Six Sigma techniques (DMAIC) methodology processes include the following steps: defined, measured, analyzed, improved, and controlled. Six Sigma works to enhance the value-adding transmutations, which happen within the process steps [40]. (Lara et al. 2022) Just-in-Time (JIT) is an organization philosophy based on inventory control referring to the mass production of goods to meet customer requirements just, in time, and quantity with the trademark of quality [41]. (Voronova et al. 2022) studied the lean warehousing design methodologies assisted tools to produce fruitful results by improving work productivity and efficiency in the overall process. Furthermore, reduced cost by eliminating waste as well as lean approach-based tools play an important role in consumer and employee satisfaction in all respects [42].

3 Observations and Discussions

This paper presents a lean-oriented strategy to organize and determine the film of activities executed on receiving and supplying goods in warehouse facilities. The warehouse activity reveals designs in detail comprising numerous activities and consumer orders that lead to enhancements in inventory design systems, warehouse layout design, and order picking procedure. The outcomes of implementing a lean concept in the warehouse arrangement showed performance-oriented results, as illustrated in Table 1.

Many researchers can put effort into the warehouses to improve order-picking accuracy problems and however, but it can be controlled and improved by analyzing goods velocity, making storage strategies, and establishing goods-picking routes. Furthermore, analyzing inventory error data and automating as much as possible for the smooth flow of the process.

Table 1 Lean implementation based results from flow processes in companies

Author's	Production unit	Lean tools	Results
(Jaca et al. 2012)	Warehousing	Continuous improvement (CI) Programmes	Enhancement in overall warehouse productivity [43]
(Dehdari 2013)	Warehousing	Lean hypothesis approach	Increase in warehouse productivity [44]
(Dujmesic et al. 2018)	Ware house	Voice Technology	Process enhancement, shorten work processes and decrease financial expenditures [45]
(Martins et al. 2020)	Cork stopper factory	Lean strategy	Increase the productivity by reducing various operational errors [46]
(Swank 2003)	Warehousing and services	Lean approach based model cell	Reduction in labour costs and reissue referable errors [47]
(Andelkovi et al. 2016)	Warehousing	Lean principles	Increase the productivity and accuracy, reduced cost [48]
(Prasetyawan and Ibrahim 2020)	Warehousing	Kanban	Using multi-criteria decision for cost and time saving [49]
(Perkumien et al. 2022)	Logistics manufacturing company	Voice system, MRP and ERP systems, ABC and JIT	Innovative solutions for the enhancement of the inventory processes [50]

4 Conclusion

From the study of review papers, it has been observed that comprehensive warehousing design methodologies have been broadly explored with contrasting research contexts, thus that an outstanding goal is far from being achieved. The following conclusions have been drawn from the summary.

- It has been observed that there are no optimization solutions for the overall design process whereby the lean-thinking-based approach and modern techniques can be used to obtain optimum design results.
- In warehouses, lean is a useful tool to enhance quality by minimizing or eliminating errors from the entire process. It is proposed that the design framework can enact research curiosity for this domain and the need for an integrated approach by applying lean strategy-based modern tools and techniques are used to address the research gaps in warehouses.
- From this study, it has been observed that the lean warehouse design is mainly related to essential components of perfection, cost-effectiveness, optimum operational performance, cleanliness, safety, and security.
- The study presented in this paper gives great ideas for warehousing optimization and sight some modern technologies that are implemented in warehouses, but only some of them are practically implemented in real-world warehouses and exploited.

References

1. Cormier G, Gunn EA (1992) A review of warehouse models. Eur J Oper Res 58:3–13
2. Hackman ST, Frazelle EH, Griffin PM, Griffin SO, Vlasta DA (2001) J Prod Anal 16(1):79–100
3. Higginson JK, Bookbinder JH, Langevin AL, Riopel D (eds) (2005) Logistics systems: design and optimization. Springer, New York. pp 67–91
4. Gupta P, Vardhan S (2016) Int J Prod Res 54(10):2976–2988
5. Singh H (2012) Intern J Adv Eng Res Stud 1(3):39–41
6. Gupta P, Vardhan S (2014) Appl Mech Mater 592:2694–2698
7. Singh H, Pandita N, Singh P, Singh K (2014) Intern J Res Appl Sci Eng Technol 2(4):206–211
8. Rouwenhorst B, Reuter B, Stockrahm V, van Houtum G, Mantel R, Zijm W (2000) Eur J Oper Res 122(3):515–533
9. Azadivar F (1986) Int J Prod Res 24(3):551–556
10. E.P. Chew, L.C Tang, European Journal of Operational Research 112, 582 (1999).
11. G. Cormier, E.A. Gunn, A review of warehouse models. European Journal of Operational Research 58, 3–13 (1992).
12. Y. Prasetyawan, N. G. Ibrahim, IOP Conf. Series: Materials Science and Engineering, 847 (2020) 012033 (doi:https://doi.org/10.1088/1757-899X/847/1/012033).
13. Gunasekaran A (1999) Int J Prod Econ 62:87–105
14. Alsmadi M, Almani A, Jerisat R, Total Qual Manag Bussiness Excel 23(4):381–396
15. I Abushaikha L Salhieh N Towers (2018) Int J Retail Distrib Manag.https://doi.org/10.1108/IJRDM-03-2018-0059
16. Schonberger RJ (1986) The Free Press, New York, NY

17. Sanchez A, Perez M (2004) Int J Serv Technol Manage 5(5):465–478
18. Malmbrandt M, Ahlstrom P (2013) Int J Oper Prod Manag 33(9):1131–1165
19. Abdi F, Shavarini S, Hoseini S (2006) J Serv Res 6:191
20. Apte U, Goh CH (2004) Int J Serv Technol Manag 5(5–6):488–506
21. Malladi S, Dominic P, Kamil A (2011) Int J Bus Inf Syst 8(3):247–268
22. Dhaouadi A, Bousselmi K, Gammoudi MM, Monnet S, Hammoudi S (2022) Data 7:113. https://doi.org/10.3390/data7080113
23. Rebelo CGS, Pereira MT, Silva FJG, Ferreira LP, Sa JC (2021) Procedia Manuf 55:471–478. https://doi.org/10.1016/j.promfg.2021.10.064
24. Namazi M (2009) Cost Manag 23(5):34
25. Shafeek H, Bahaitham H, Soltan H (2018) J Comput Theor Nanosci 15(6/7):1814–1817
26. Raghuram P, Arjunan MK (2022) Int J Product Perform Manag 71(6):2410–2431
27. Metha VB, Dave P (2020) Int J Adv Eng Manag 2(4):421–436
28. Singh J, Singh H (2009) IUP J Oper Manag 8(2):51–72
29. Jaffar A, Kasolang S, Ghaffar ZA, Mohamad NS, Khairul M, Mohamad F (2015) J Teknol (Sci Eng) 76(6):19–23
30. Kumar R, Dwivedi RK, Verma A (2016) Indian J Eng, 13(33), 362–370
31. Carvalho CP, Gonçalves LWN, Silva MB (2019) Int J Prod Manag Eng 7(1):1–12. https://doi.org/10.4995/ijpme.2019.8607
32. Amrina E, Putri NT, Anjani DM (2019) IOP conf. series: materials science and engineering, p 528. https://doi.org/10.1088/1757-899X/528/1/012051
33. Patel AS, Patel KM (2021) IOP conf. series: materials science and engineering, p 1070. https://doi.org/10.1088/1757-899X/1070/1/012133
34. Wolniak R, Zasadzien BS, Kwiecien AG (2018) MATEC web of conferences 183. https://doi.org/10.1051/matecconf/201818301001
35. Grzelczak A, Lewandowska KW (2016) Arab J Bus Manag Rev 6(3). https://doi.org/10.4172/2223-5833.1000216
36. Vargas AR, Soto KCA, Gutierrez TC, Ravelo G (2018) Appl Sci 8
37. Dalgobind M, Anjani K (2008) J Ind Eng Anagement 1(2):16–53. https://doi.org/10.3926/jiem.v1n2.p16-53
38. Bilalis N, Scroubelos G, Antoniadis A, Emiris D, Koulouriotis D (2002) Int J Prod Res 40(15):3575–3588
39. Rahman NA, Sharif SM, Esa MM (2013) Procedia Econ Financ 7:174–180. https://doi.org/10.1016/S2212-5671(13)00232-3
40. Jiju A, Ronald S, Roger H, Ton VDW (2017). Int J Qual Reliab Manag. https://doi.org/10.1108/IJQRM-03-2016-0035
41. Lara AC, Menegon EMP, Sehnem S, Kuzma E (2022) A meta-analysis. Gest Prod 29. https://doi.org/10.1590/1806-9649-2022v29e9021
42. O. Voronova, Transportation Research Procedia, 63, 919–928 (2022).
43. Jaca C, Viles E, Mateo R, Santos J (2012) TQM J 24(2):142–154
44. Dehdari P (2014) KIT Scientific Publishing. pp 1–191. https://doi.org/10.5445/KSP/1000036424
45. Dujmesic N, Bajor I, Rozic T (2018) Technical Gazette 25:1227–1233
46. Martins RM, Pereira T, Ferreira LP, Sa JC, Silva FJG (2020) Procedia Manuf 51:1723–1729
47. Swank C (2003) Harv Bus Rev 81:138
48. Andelkovi, Radosavljevi M, Pani DS (2016) Econ Themes 54(4):517–534
49. Prasetyawan Y, Ibrahim NG (2020) IOP conf. series: materials science and engineering, p 847
50. Perkumien D, Ratautait K, Pranskunien R (2022) Sustainability 14

Implication of MFO for Control of 3-link Robotic Manipulator Used for Casting Process

Check for updates

Mahendra Kumar Jangid, Sunil Kumar, and Jagtar Singh

Abstract The inverse kinematics of the 3-link robotic manipulator was solved utilizing the moth-flame optimization (MFO) algorithm in this study. The outcomes were compared to those obtained using other optimization methods, including the grey wolf optimization (GWO) algorithm, particle swarm optimization (PSO) algorithm and whale optimization algorithm (WOA). First, the transformation matrices and D-H values of the robotic arm are generated. The end-effector position equations are then developed using the general transformation matrix. Using the MFO, PSO, GWO, and WOA, this robotic manipulator's end-effector position in the working area is estimated. A fitness function is used to estimate the position error or the distance between the current position and the desired place. By utilizing the fitness function to minimize the position error, the inverse kinematics solutions were produced. These algorithms were examined in this study, two distinct examples were used. Error in position and time to solve were estimated in Case-I for one place in the workspace, whereas Case-II estimated error in position and time to solve for 20 arbitrarily chosen areas of the workspace. By comparing it to case-I, case-II demonstrates the superiority of the MFO method over additional optimization techniques (PSO, GWO, and WOA). The MFO algorithm performs significantly better than PSO, GWO, and WOA algorithms with regard to errors in position error and time to solve, according to the results.

M. K. Jangid (✉) · S. Kumar · J. Singh
Department of Mechanical Engineering, Sant Longowal Institute of Engineering and Technology, Longowal 148106, India
e-mail: mahendra.jangid0007@gmail.com

S. Kumar
e-mail: sunilkumar@sliet.ac.in

J. Singh
e-mail: jagtarsliet@gmail.com

A. Sachdeva et al. (eds.), *Recent Advances in Operations Management and Optimization*, Lecture Notes in Mechanical Engineering,
https://doi.org/10.1007/978-981-99-7445-0_15

Abbreviations

MFO	Moth-flame Optimization
PSO	Particle Swarm Optimization
GWO	Grey Wolf Optimization
WOA	Whale Optimization Algorithm
D-H	Denavit Hartenberg

1 Introduction

The need for robotic manipulators is growing, day-by-day in various industries like healthcare sector, agriculture, warehouse, automobile manufacturing industry, etc. The robotic manipulator is more efficient, accurate and stable than human for the fast manufacturing process and to replace boring stuff that is performed by human beings [1]. Inverse kinematics solution of the robotic manipulator gets complicated when the number of degrees of freedom (D.O.F.) increases. A major difficulty in robotics is inverse kinematics, which is traditionally involved in issues like path planning, motion creation, and trajectory optimization. Inverse kinematics is necessary for articulated robots in particular to generate joint motions that correspond to a given target location. Because arms and legs are often part of articulating systems, they are crucial in humanoid robotics. Precise positioning of the end-effector and accurate joint angles are very necessary when programming the robot like in the semiconductor manufacturing industry [2]. Today's businesses depend heavily on robot arms because of their accuracy in high-speed manufacturing. Solving inverse kinematics is one of the most difficult issues faced by industrial robots. Calculating the angles' values that are connected to the desired Cartesian position is the focus of the Inverse Kinematic Problem. Inverse kinematic problems can now be solved more quickly and with sufficient solutions utilizing soft computing techniques as opposed to the more conventional numerical, geometric, and algebraic approaches. In the past ten years, numerous studies have been done on the use of soft computing algorithms to solve the inverse kinematics of robotic manipulators. Çavdar et al. [3] have proposed a new heuristic approach based on the modified Artificial Bee Colony (ABC) algorithm for the inverse kinematics solution of a Puma 560 robotic arm. The authors minimize the distance between the target position and the actual position using a fitness function and get the joint angles by modified ABC algorithm and compare the obtained results using other optimization algorithms such as Particle Swarm Optimization (PSO) and Harmony Search (HS). The authors concluded that modified ABC gives good results in terms of error in position and time to solve in contrast with PSO and HS. Rokbani and Alimi [4] proposed an IK-PSO algorithm for the solutions for inverse kinematics on 2 DOF planer robotic manipulator and make. This study compares how well different PSO variations perform while solving inverse kinematics. An articulated system with double links employs inverse kinematics utilizing PSO. There is a statistical analysis to evaluate the main PSO variations' convergence and comparative performances, The restriction factor PSO, linear declining weight, inertia factor PSO, and two easier PSO versions, when used to solve IK.

Huang and Hsu [5] proposed a biologically motivated Deoxyribonucleic Acid (DNA) algorithm for the solution of inverse kinematics solution of a 5 DOF robot manipulator. The simulation results are also compared with the optimization algorithm Genetic Algorithm (GA) and PSO algorithm. The authors concluded that the biologically inspired deoxyribonucleic acid (DNA) algorithm shows effective findings in comparison to GA and PSO in terms of position error in x, y and z directions of the end-effector. Mustafa et al. [6] use Artificial Neural Network (ANN) to solve the inverse kinematic (joint angles) of a 3-DOF plasma cutting robot. The authors also compare the results obtained from the MSC.ADAMS software. The authors finding concluded that ANN produced good results in terms of simplicity and accuracy. Mustafa and Kerim [7] designed and developed a 4-DOF robotic manipulator for pick and place tasks and conducted a study on inverse kinematics solutions using four soft computing methods. The algorithms are the quantum Particle Swarm Optimization (QPSO) algorithm, the gravitational search method, the PSO algorithm, and the GA algorithm (GSA). All of these algorithms underwent comparative analysis after being put to the test in two different circumstances. The authors concluded that the QPSO algorithm shows beneficial outcomes in terms of execution and position time for both scenarios as compared to the PSO algorithm, GA, and GSA. Ze Fan Cai et al. [8] studied on inverse kinematic solution of a 6-joint robot using simulated annealing particle swarm optimization (SA-PSO) and compared it to PSO. The authors concluded that SA-PSO provides efficient outcomes in comparison to PSO. Qun and Jiajun [9] studied on inverse kinematics solution of a 6-DOF robot with offset-wrist using Adaboost Neural Network. The author concluded that the Adaboost Neural Network gives good results as compared to previous studies in the literature. Zhiyu Zhou et al. [10] suggest a clever approach built on extensive learning machines and genetic algorithms with successive mutation to find the inverse kinematics of a 6-DOF robotic manipulator. This approach was created to reduce calculation without compromising end effector precision. In the suggested technique, computation of the initial inverse kinematics solution by an advanced learning machine initially is optimized using a new genetic algorithm based on sequential mutation. The suggested advanced learning machine and consecutive genetic method for mutagenesis can significantly increase time effectiveness while assuring accuracy at the end-effector, according to the experimental results. Ahmed et al. [11] This paper compares various soft-computing-based approaches to the inverse kinematics problem, including Artificial Neural Networks, Adaptive Neuro Fuzzy Inference Systems, and Genetic Algorithms. Both ANN and ANFIS are able to perform better than alternative approaches with the aid of a proposed mechanism termed minimal error function. The simulation results were validated by analysis of the experimental data utilizing a 5DOF robot arm.

Dereli and Köker [12] conducted a study on the inverse kinematic solution of a 7-DOF redundant robot manipulator using Firefly Algorithm (FA). The author also makes a comparison to other optimization algorithms like PSO and Artificial Bee Colony (ABC) algorithm. The authors came to the conclusion that FA outperforms PSO and ABC in terms of the error in position and time to solve. Zhang and Xiao [13] studied on the inverse kinematic solution of a 7-DOF robotic manipulator using a chaotic and Parallelized Artificial Bee Colony (CPABC) algorithm. The author concluded that CPABC gives successful results with respect to errors in location and time to solve as compared to another optimization algorithm. Samta et al. [14] In this study, three alternative optimization algorithms-PSO, 'segmented particle swarm optimization' (SPSO), and 'modified segmented particle swarm optimization' (MSPSO)-reused to maximize the weights and biases of neural networks. The results are then compared. Additionally, MSPSO outperforms the other two optimization methods in a comparison based on two factors, namely fitness function and regression. Heng and Chong [15] Using two serial robotic manipulators, the efficacy of 'adaptive particle swarm optimization' (APSO) was assessed in this study. In contrast with APSO, a variety of potent PSO variants that were created utilizing diverse methods were chosen. The results of the experiments demonstrate that the prescribed exercise program that works with APSO may precisely and successfully deal with the problem of inverse kinematics in various DOF manipulators.

Yiyang et al. [16] to maintain the particle update rate to adjust to every step in the optimization procedure, a complicated utilizing complex inertia mass the idea of resemblance is included, making the to find out more dependable. The problem of the most effective remedy locally is also addressed. Numerous populations are produced in order to conduct simultaneous optimization searches, as well as the immigrant operator is recommended to increase each iteration's component population's diversity. The study's findings, which were validated using the Comau NJ-220 robot, prove that the suggested augmented PSO uses better algorithm robustness for problems with inverse solutions for broad robot kinematics and has the potential to dramatically improve convergence precision and speed. This strategy presents a novel remedy for the inverse kinematics problem of robots and provides a superior solid and useful planning of robot motion based on kinematics. Rahkar [17] proposed a novel metaheuristic optimization method dubbed BRO (battle royale optimization). The video game genre known as "battle royale" served as the model for the suggested tactic. Each individual is shown as a soldier or player in the population-based algorithm BRO who seeks to move to the greatest (safest) neighborhood and eventually live. The recommended method has been tested against six recently developed optimization algorithms as well as the well-known PSO technique for the 6-DOF PUMA 560 robot arm's inverse kinematics solution. The results of the experiments show that the recommended algorithm is a useful method that produces favorable and competitive outcomes in terms of reliability and uniformity. In Sandi Baressi et al. [18], a feed-forward artificial neural network called a multilayer perceptron (MLP) is trained to calculate the inverse kinematics of a robotic manipulator. First, the direct kinematics of a robotic manipulator are derived using the D–H method. The predicted homogenous transformation matrices are then used to create a dataset of 15,000 points. Following the exercise, 10,240 unique mixtures of hyperparameter values on several MLPs, the best is then chosen. The MLPs' underlying topologies include the greatest outcomes supplied, and the R2 and mean absolute error metrics are used to evaluate each trained MLP. The regression of the last joint fared inferior to the first five joints in terms of quality due to the design of the robotic manipulator (The error rate was around 0.1 percent.). One more recent similar study using Ant Lion Optimization (ALO) was conducted by M. K. Jangid et. al. [22]

This research's objective is to use Moth-Flame Optimization (MFO) techniques to address the inverse kinematics of a whole new 3-link robotic arm used in casting process applications. Error in position and time to solve were used to test this approach, along with comparisons to additional optimization techniques including PSO, GWO, and WOA. I and II are two cases were compared to determine the efficacy and efficiency of these algorithms.

2 Robotic Manipulator Kinematic Modelling

SolidWorks software was used to design and create a 3-link robotic manipulator for the purpose of casting in a brass pellet production business, as shown in Fig. 1a, b. Stepper motors with epicyclic gear trains were employed as the acuter for joints 1, 2, and 3. The stepper motor drivers with a 12 V and 5-Amp power supply were used to operate/control these motors. Two tiny stepper motors are also used to drive an end-effector with 2 degrees of freedom that rotates and turns a small crucible.

This robotic manipulator's end-effector position is determined using forward kinematics. By applying Denavit–Hartenberg (D–H) characteristics, frames are assigned to robotic manipulators as the first action in the calculation of 'forward kinematics'. Figure 2a of this study displays the assignment of frames on a 3-link robotic arm and Fig. 2b displays the end-effector. As in Table 1, the frame assignment and robot dimensions are now used to generate the D–H parameters table. Then, by

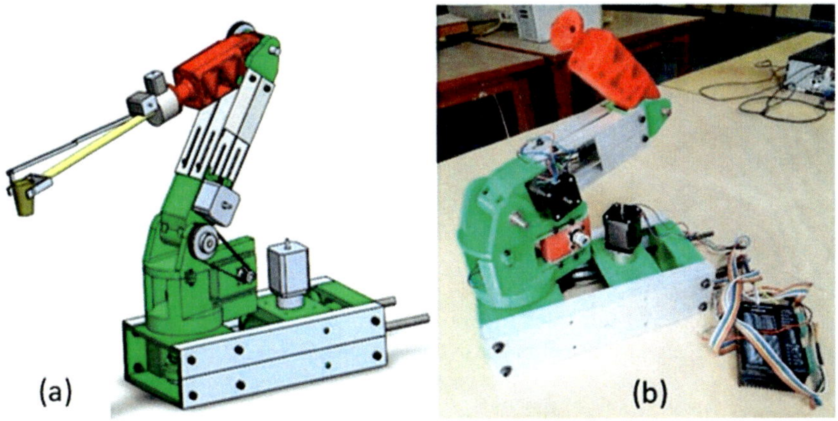

Fig. 1 3-link robotic manipulator **a** CAD design **b** manufactured robot

inserting values from D–H values into Eq. (1), matrix transformations for forward kinematics from Eq. (2) to Eq. (5) are created. The end-effector position equations are Eqs. (6), (7), and (8) for locations of the end-effectors represented by P_X, P_Y, and P_Z, respectively, and are produced by the general transformation matrix Eq. (5).

$$T_i^{i-1} = \begin{bmatrix} C\theta_i & -S\theta_i C\alpha_i & S\theta_i S\alpha_i & a_i C\theta_i \\ S\theta_i & C\theta_i C\alpha_i & -C\theta_i S\alpha_i & a_i S\theta_i \\ 0 & S\alpha_i & C\alpha_i & d_i \\ 0 & 0 & 0 & 1 \end{bmatrix} \tag{1}$$

$$T_1^0 = \begin{bmatrix} C_1 & 0 & S_1 & 0 \\ S_1 & 0 & -C_1 & 0 \\ 0 & 1 & 0 & d_1 \\ 0 & 0 & 0 & 1 \end{bmatrix} \quad T_2^1 = \begin{bmatrix} C_1 & -S_2 & 0 & a_2 \cdot C_2 \\ S_2 & C_2 & 0 & a_2 \cdot S_2 \\ 0 & 0 & 1 & 0 \\ 0 & 0 & 0 & 1 \end{bmatrix} \tag{2}$$

$$T_3^2 = \begin{bmatrix} S_3 & 0 & -C_3 & a_3 \cdot S_3 \\ C_3 & 0 & S_3 & a_3 \cdot C_3 \\ 0 & -1 & 0 & 0 \\ 0 & 0 & 0 & 1 \end{bmatrix} \quad T_4^3 = \begin{bmatrix} C_4 & -S_4 & 0 & 0 \\ S_4 & C_4 & 0 & 0 \\ 0 & 0 & 1 & d_4 \\ 0 & 0 & 0 & 1 \end{bmatrix} \tag{3}$$

$$T_5^4 = \begin{bmatrix} C_5 & 0 & S_5 & 0 \\ S_5 & 0 & -C_5 & 0 \\ 0 & 1 & 0 & 0 \\ 0 & 0 & 0 & 1 \end{bmatrix} \tag{4}$$

$$T_5^0 = T_1^0 \cdot T_2^1 \cdot T_3^2 \cdot T_4^3 \cdot T_5^4 = \begin{bmatrix} n_x & s_x & a_x & P_x \\ n_y & s_y & a_y & P_y \\ n_z & s_z & a_z & P_z \\ 0 & 0 & 0 & 1 \end{bmatrix} \tag{5}$$

$$P_x = -C_1 \cdot (d_4 \cdot C_{23} - a_2 \cdot C_2 + a_3 S_{23}) \tag{6}$$

$$P_y = -S_1 \cdot (d_4 \cdot C_{23} - a_2 \cdot C_2 + a_3 S_{23}) \tag{7}$$

$$P_z = d_1 + a_2 \cdot S_2 + a_3 \cdot C_{23} - d_4 \cdot S_{23} \tag{8}$$

where: $C_1 = \text{Cos}(\theta_1)$, $S_1 = \text{Sin}(\theta_2)$............etc. And $C_{23} = \text{Cos}(\theta_2 - \theta_3)$, $S_{23} = \text{Sin}(\theta_2 - \theta_3)$.

θ_4 and θ_5 have a fixed value of zero degrees. These two values do not contribute any effect on the end-effector position but change the orientation of the end-effector.

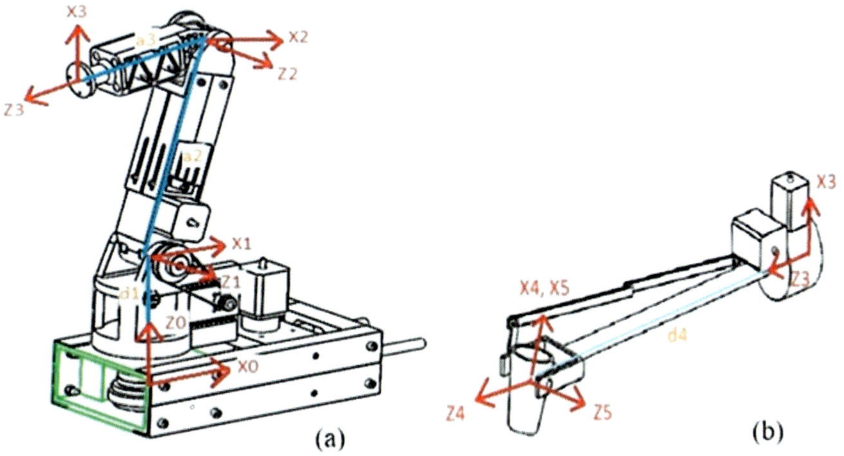

Fig. 2 D–H assignments of frames of **a** 3-link robotic manipulator **b** end-effector

Fig. 3 Moth's flying towards light source on a spiral path [20]

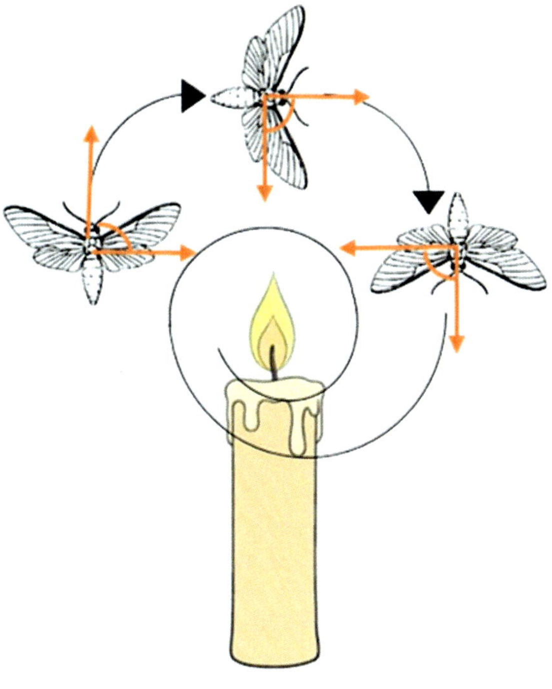

Table 1 D–H parameters table

Link$i$	$a_i(m)$	$\alpha_i(°)$	$d_i(m)$	$\theta_i(°)$(Range)
1	0	90°	d1 $= 0.19$	$\theta_1(-135to135)$
2	$a_2 = 0.25$	0	0	$\theta_2(-90to90)$
3	$a_3 = 0.16$	90°	0	$\theta_3 - 90(-90to90)$
4	0	0	d4 $= 0.26$	$\theta_4(-180to180)$
5	0	90°	0	$\theta_5(-45to45)$

3 Moth-Flame Optimization

The Moth-Flame Optimization (MFO) is a brand-new optimization technique that draws inspiration from nature and is based on how moths navigate at night while retaining a constant relation to the moon. Mirjalili introduces it in the beginning [19] and conducts experiments on seven engineering problems and 29 benchmark functions, the author concluded that MFO shows effective results as compared to other nature-inspired optimization algorithms. Here are the steps for MFO.

3.1 Mapping and Navigation of Moths in MFO

The transverse orientation, a moth's night time navigation technique, serves as the foundation for MFO, an optimization algorithm inspired by nature. If they maintain a steady angle with the Moon, moths can fly huge dimensions of straight lines. Moths attempt to keep a similar angle to artificial lights when they come into contact with them, but due to the near proximity, they are caught in a spiral route, as shown in Fig. 4. Each moth has a unique fitness function value, and in the solution space of the optimization issue, the MFO assigns moths to various solutions. Each moth also possesses a flame that it uses to store the best solution it has ever discovered. In each iteration, the moths fly in a spiraling motion toward their fires to update their positions and search the solution space. In the MFO solution space, moth positions are first initialized at random.

The moths' computed fitness values are the highest individual fitness values yet. For each moth, the flame designates the ideal unique position. The positions of the moths get updated using a spiral as a basis motion work toward their greatest various positions, which are displayed by a flame, whose positions are updated in the subsequent iteration with the new greatest individual positions. Until the ending criteria are satisfied, the MFO algorithm keeps updating the moth and flame locations and creating new locations. The MFO's flowchart is shown in Fig. 4.

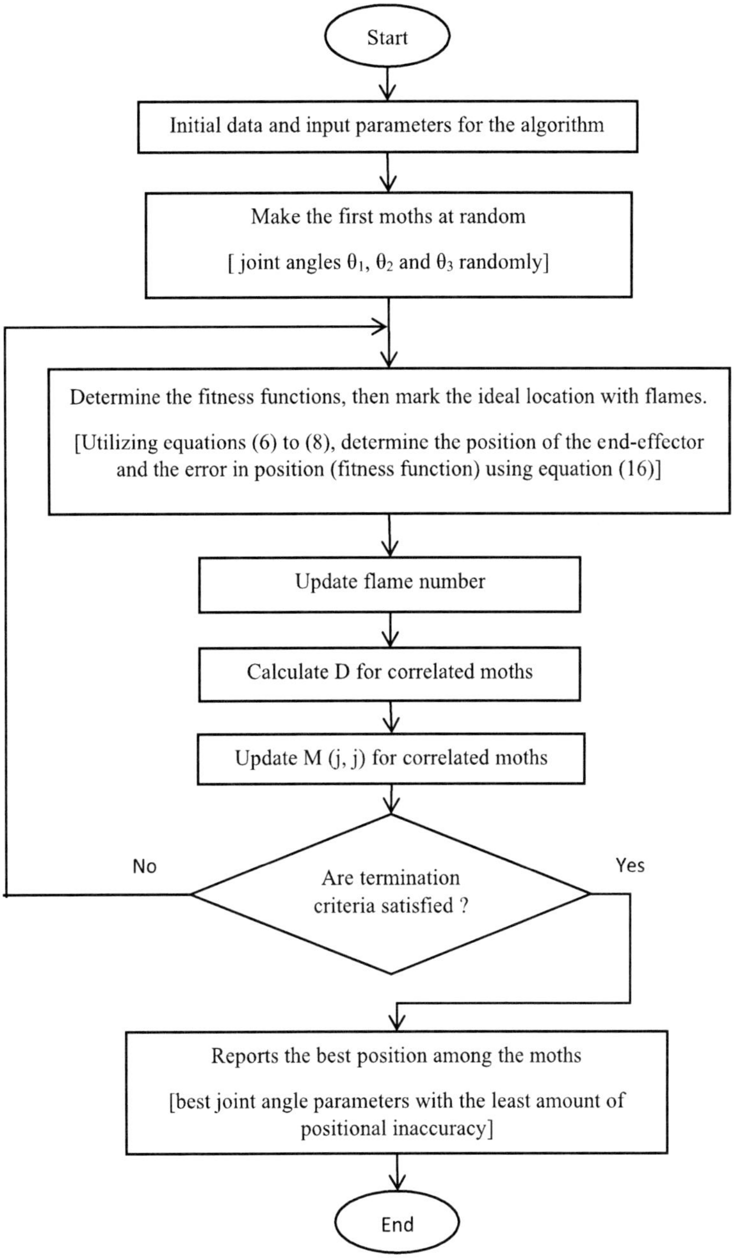

Fig. 4 MFO flowchart

3.2 Initial Population of Moth's

When employing MFO, each moth is taken into account as having a place in a D-dimensional optimal solution. Equation (9) can be used to express the collection of moths, where n is the total of moths, where d is the overall count of parameters in the problem space. As indicated in the equation, the moths' associated fitness function values are set up in an array Eq. (9).

$$M = \begin{bmatrix} m_{1,1} & m_{1,2} & \cdots\cdots & m_{1,d} \\ . & . & . & . & . \\ . & . & . & . & . \\ . & . & . & . & . \\ m_{n,1} & m_{n,2} & \cdots\cdots & m_{n,d} \end{bmatrix} \text{OM} = \begin{bmatrix} \text{OM}_1 \\ \text{OM}_2 \\ . \\ . \\ \text{OM}_n \end{bmatrix} \tag{9}$$

Two more elements of the flame matrix are made up of MFO., which represents a fitness function vector that corresponds to the flames in the D-dimensional space, which is possibly conveyed mathematically in Eq. (10).

$$F = \begin{bmatrix} F_{1,1} & F_{1,2} & \cdots\cdots & F_{1,d} \\ . & . & . & . & . \\ . & . & . & . & . \\ . & . & . & . & . \\ F_{n,1} & F_{n,2} & \cdots\cdots & F_{n,d} \end{bmatrix} \text{OF} = \begin{bmatrix} \text{OF}_1 \\ \text{OF}_2 \\ . \\ . \\ \text{OF}_n \end{bmatrix} \tag{10}$$

Moths search the space of possible solutions throughout each iteration of MFO to locate the best one, and the flames reflect the ideal response that every moth has discovered. Alternatively, every moth looks about its flame until it discovers a better response. The flame's position is then updated.

3.3 Changing the Moths' Location

The three functions used by MFO to the moths around in the area of the solution (S), initialize the moths' random positions (I), and end the search operation (O) are as follows:

$$\text{MFO} = (I, S, O) \tag{11}$$

The initialization of the moth positions in the solution space can be done using any random distribution. Equation can be used to express the I function's implementation (12).

$$M(i,j) = \left[ub(i) - lb(j)\right] \times rand() + lb(i) \tag{12}$$

where the arrays ub and lb, which stand for the upper and lower limits of variables, respectively. A logarithmic spiral is used to simulate the motion of the number of moths in the optimal way out built around their transverse orientation under the criteria listed below:

- The spiral's entry point ought to be the moth.
- The spiral's exit point ought to be where the flame is located.
- Spiral range fluctuation shouldn't go beyond the search area.

Consequently, the movement's P function is defined as in Eq. (13).

$$S\left(M_i, F_j\right) = D_i.e^{bt}.\cos(2\pi t) + F_j \tag{13}$$

where Di = the distance between ith moth and the jth flame, which is illustrated in Eq. (13), and $b = a$ factor that helps define the curve of the spiral logarithm, falling between [−1, 1].

$$D_i = \left|F_j - M_i\right| \tag{14}$$

Exploration and use of the solution space are ensured by the moth's spiralizing motion around the flame. In each iteration, the top solutions (flames) are sorted to prevent the moths from being stuck in local optima, and Each moth uses the OF and OM matrices to move around the flame that corresponds to it. The initial moth then circles the ideal response discovered; the most recent month circles the worst course of action discovered.

3.4 Updating the Flames Count

Equation (15) is used to reduce the quantity of flames and, as a result, the ideal solution just attracts moths about in the latter stages of the algorithm, improving the MFO method's performance.

$$F.N. = rand\left(N - l \times \frac{N - l}{T}\right) \tag{15}$$

where l is the current iteration count, N is the maximum number of flames, and T is the maximum iteration count. Exploration and use of the solution space are balanced by the decline in the number of flames.

3.5 Termination Requirements

The algorithm is terminated based on the termination criterion. A good termination condition must be chosen in order for the method to converge correctly. Common MFO termination criteria include the number of repetitions, the extent of the improvement, and the duration.

4 Fitness Function

The main goal of this study is to acquire the robotic manipulator joint angles for the stated path/position with the least amount of position error possible. As shown in Fig. 5, during the casting process for metal liquid pours, a robotic manipulator end-effector follows a specified trajectory. By reducing position error—the distinction between the goal position (P_2) and the precise location (P_1)-the precise trajectory is obtained. M.K. Jangid et al. [21] carried out a similar investigation on trajectory tracking optimization during the pouring of liquid metal in casting.

$$error = \sqrt{(X_2 - X_1)^2 + (Y_2 - Y_1)^2 + (Z_2 - Z_1)^2} \tag{16}$$

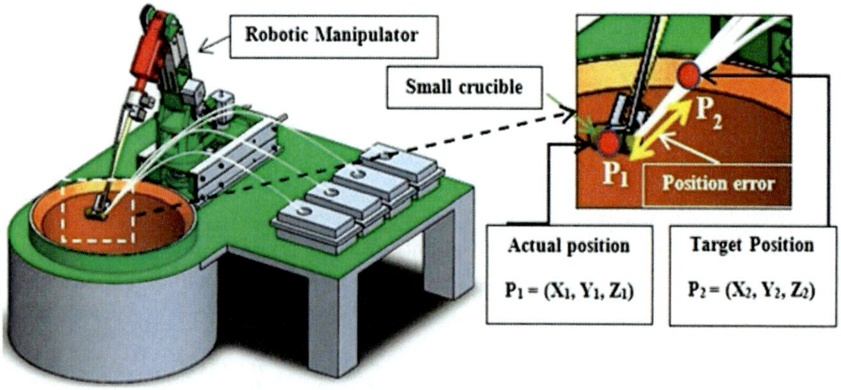

Fig. 5 Robotic manipulator end-effector position inaccuracy presented in a CAD model while metal is being poured from a tiny crucible

5 Simulation Results

In this work, the MFO algorithm was tested for inverse kinematics solutions in comparison to alternative optimization techniques including PSO, GWO, and WOA. All algorithms were developed in MATLAB version 2016a and performed on a computer with an Intel(R), 'Core (TM) i3-7020U CPU' operating at 2.30 GHz as part of the study's two cases, which were compared at the end. In both cases, errors in position and computation time are prioritized.

5.1 *Case I*

One point is selected from the robotic manipulator's work in the first scenario area Fig. 6a as $[P_X, P_Y, P_Z] = [0.1226, 0.0393, 0.5628]$ These numbers are attained by substituting $[\theta_1, \theta_2, \theta_3]$ values for [60°, 45°, 45°] in Eq. 6 through 8.

Table 2 displays the ideal joint angle values for Case-I as determined by several optimization procedures. These figures are attained by choosing the optimal minimum error in position and minimum time to solve after 10 iterations of the procedure.

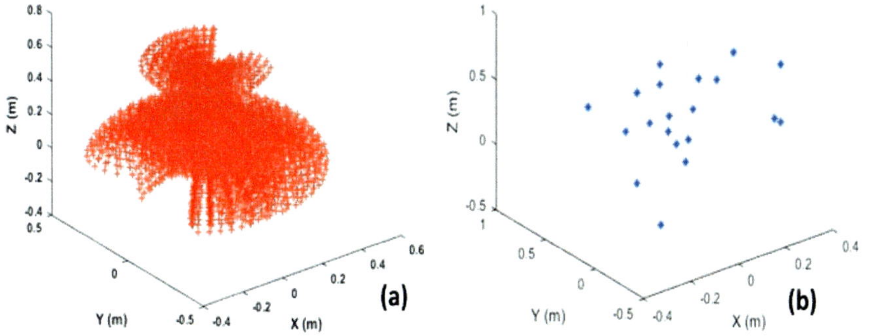

Fig. 6 **a** Workspace of the robotic manipulator in three dimensions **b** 20 randomly chosen locations in the working area

Table 2 Values for the joint angles produced by optimization techniques

Angles	Range (in degree)	Manual values	PSO	WOA	GWO	MFO
θ_1	−135° to 135°	60°	22.3098	113.4071	−122.1671	−103.3629
θ_2	−90° to 90°	45°	−17.3056	1.213757	−61.81894	−55.5312
θ_3	−90° to 90°	45°	−2.5887	59.87798	13.57607	76.41571

Table 3 displays the end-effector position's ideal values and its error in position in comparison to additional optimization techniques. As shown in Fig. 7, the MFO method produces effective outcomes when compared to other algorithms in terms of position inaccuracy. These solutions are created by choosing a different particle size for each algorithm iteration, which is done after the algorithm has been run ten times.

Now, the recommended particle size for outcomes is provided in Table 3. The most iterations allowed for any of these algorithms is 500. At approximately 200 iterations and a solution duration of almost 14 s, the PSO algorithm provides an efficient solution.

Utilizing the GWO results in a resolution after iterations close to 400 and a time to solve close to two seconds. The WOA delivers the response at approximately 280 iterations and nearly 1.3 s of solution time, whereas the last MFO provides the answer at nearly 260 iterations and nearly 0.8 s of solution time.

Therefore, when compared to PSO, GWO, and WOA, MFO produces very good results with regard to position inaccuracy and response time. Comparing the WOA algorithm to the PSO and GWO algorithms, it produces good results. Additionally, it has been discovered whenever the quantity of particles in a WOA increases, the position inaccuracy decreases and the time to solutions lengthens. Additionally, it has been discovered that while the results of position/location inaccuracy do not change as much as the quantity of agents raised in MFO, the amount of time it takes to solve the problem does.

Table 3 End-effector position determined using optimization techniques

End-effector position (m)	Manually entered values	PSO	WOA	GWO	MFO
P_X	0.1226	0.1262	0.12174	0.1198	0.1225
P_Y	0.0393	0.0417	0.03897	0.0445	0.039195
P_Z	0.5628	0.5699	0.56369	0.5614	0.5626998
Error (m)		0.00836	0.001271	0.006882	7.653e−16

Table 4 Comparing MFO to alternative optimization techniques

	Particle size	Max. iteration	Solution iteration	Error in position (m)	Time to solve (s)
PSO	300	500	229	0.00836	14.35
WOA	150	500	280	0.001271	1.34
GWO	150	500	404	0.006882	2.02
MFO	30	500	263	7.653e−16	0.85

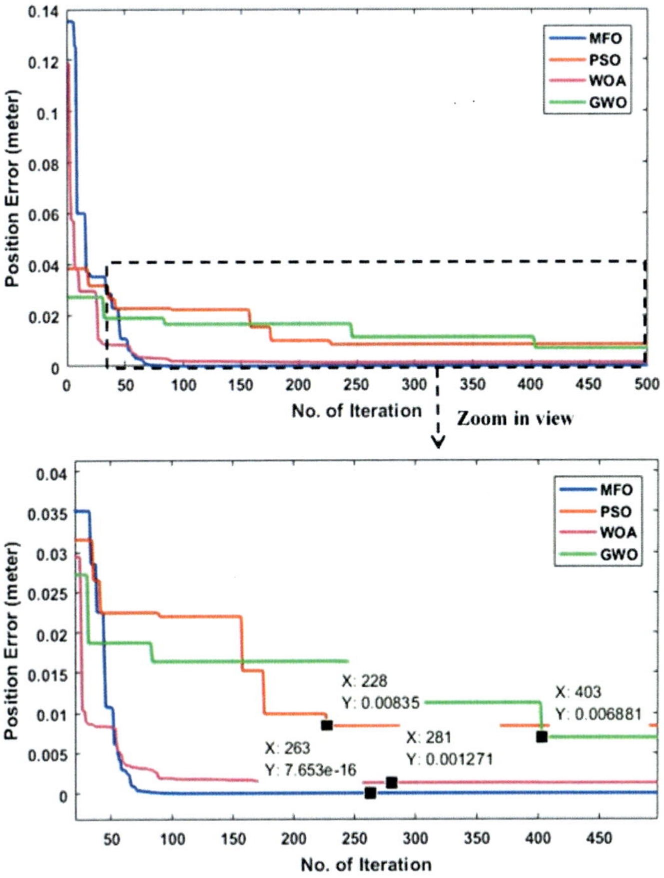

Fig. 7 For scenario I, an error in position versus iteration graph

5.2 *Case II*

In this instance, as shown graphically in Fig. 6b, 20 randomly chosen locations from the workspace Fig. 6a were chosen. The following are utilized to compare how well the algorithms work employed in case I since the randomly chosen point in question provides several numbers for error in position and time to solve than the one before it.

Therefore, as we can see in Fig. 8a, MFO provides significantly better results for position inaccuracy than other methods. As opposed to PSO and GWO, the WOA algorithm also produces some impressive outcomes.

Figure 8b displays the time to solve the graph for 20 randomly selected points. and as can be seen, MFO provides a relatively short minimum solution time in comparison to other methods. In this investigation, PSO provides the slowest solution time.

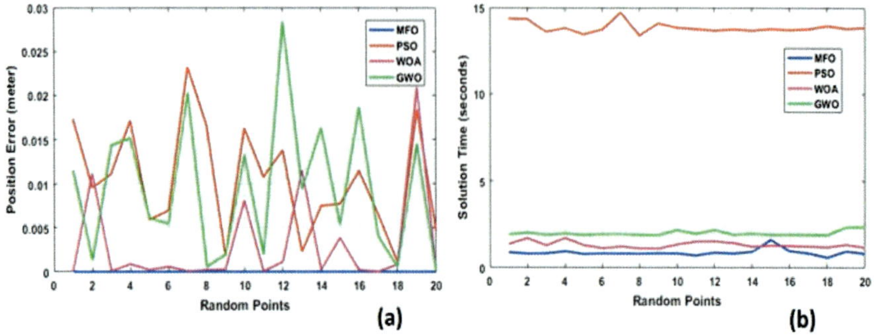

Fig. 8 **a** Error in position for 20 randomly chosen locations in the workspace **b** time to solve for 20 randomly chosen locations in the workspace

Table 5 contains all of the outcomes for Case II. It is evident that, when compared to PSO, WOA, and GWO, MFO delivers efficient results with regard to position inaccuracy and response time.

The study's main conclusions are listed below.

- For case I, the MFO provides a position error that is 100% less than that of the PSO, GWO, and WOA.
- For case I, the MFO takes 94.07%, 57.71%, and 36.57% less time to solve the problem than the PSO, GWO, and WOA.
- For MFO, PSO, GWO, and WOA, the iteration of the solution for Case-I is around 260, 200, 400, and 280, respectively.
- As opposed to PSO, GWO, and WOA, the MFO provides 100% reduced position error for case II.
- As opposed to PSO, GWO, and WOA, respectively, the MFO provides 93.68, 56.42, and 34.58% less solution time for Case II.

Results show that MFO considerably outperforms alternative methods utilized in this investigation in terms of response time and positional accuracy. MFO can be applied to solutions to inverse kinematic problems to reduce position inaccuracy and speed up solution times.

A few studies that compared the results of inverse kinematics solutions in terms of position inaccuracy and solution duration are provided in Table 6.

Table 5 ALO and other optimization techniques are compared for randomly selected elements

	Particle size	Max. iteration	Error in position (m) (avg. of 20)	Time to solve (s) (avg. of 20)
PSO	300	500	0.010495	13.84094
WOA	150	500	0.00301	1.336704
GWO	150	500	0.009462	2.006824
MFO	2	500	9.89e−16	0.87452

Table 6 Comparison with other literary studies

Author	Robotic arm	Technique employed	Error in position (m)	Time to solve (s)
Çavdar et al. [3]	PUMA560	Modified ABC	6.31059E−13	0.028618442
Rokbani and Alimi [4]	2-DOF	IK-PSO		
Huang and Hsu [5]	5-DOF	Biologically inspired deoxyribonucleic acid (DNA) algorithm	$X = 30–29.96 =$ 0.04 cm = 4E−04 m Y = 2E−04 m Z = 1E−04 m	
Mustafa et al. [6]	3-DOF plasma cutting robot	ANN	Path 1 0.26% joint 2 Path 2 1.15% joint 2	
Mustafa and Kerim [7]	4-DOF	QPSO	Scenario 1 6.95E−06 (avg.)	1.65
			Scenario 2 6.51E−06 (avg.)	1.96
Ze Fan Cai et al. [8]	6-DOF	SA-PSO	1E−05	
Qun and Jiajun [9]	6-DOF with offset-wrist	Adaboost based feed forward neural network	2.67E−06	3E-04
Dereli and Köker [12]	7-DOF	FA	6.53E−06	0.9204
Zhang and Xiao [13]	7-DOF	CPABC	2.576138E−06 (Best)	0.03592
Samta et al. [14]	3-DOF (planer)	MSPSO	6.18E−09 (theta-1) (fitness value)	
Heng and Chong [15]	6-DOF and 7-DOF	APSO	1.19E−08 (Best)	Less than 2t
Yiyang et al. [16]	Comau NJ-220 robot	Improved PSO	8xE−04 (mm) (best)	
Rahkar [17]	PUMA 560	BRO	1.89xE−07 (mean at pop. 300)	
Present study	3-DOF	MFO	9.89E−16	0.87452

6 Conclusion

In this investigation, inverse kinematics calculations of a serial robot manipulator with three degrees of freedom have been done through simulations in order to confirm the precision and effectiveness of the MFO algorithm. Heuristic functions are well suited for usage in the inverse kinematics process because of their complexity and a high degree of difficulty. Due to the indefinite size of the solution set for redundant manipulators, each method's joint angle results are unique. The results obtained with MFO are contrasted with those obtained using PSO, GWO, and WOA algorithms in order to determine the algorithm's effectiveness. Two distinct scenarios have been used in experiments. The second scenario has confirmed the first case's predictions made based on a single manually chosen point, proving the stability of the algorithm. MFO performs far superior over competing algorithms in tests, both in regard to the time to solve and position precision. In terms of calculation time and position inaccuracy, MFO is a simple tool to utilize in inverse kinematics.

References

1. Wu W, Rao SS (2007) Uncertainty analysis and allocation of joint tolerances in robot manipulators based on interval analysis. Reliab Eng Syst Saf 92(1):54–64
2. Hasan AT, Hamouda AMS, Ismail N, Al-Assadi HMAA (2006) An adaptive-learning algorithm to solve the inverse kinematics problem of a 6 D.O.F serial robot manipulator. Adv Eng Softw 37(7):432–438
3. Çavdar T, Mohammad M, Milani RA (2013) A new heuristic approach for inverse kinematics of robot arms. Adv Sci Lett 19(1):329–333
4. Rokbani N, Alimi AM (2013) Inverse kinematics using particle swarm optimization, a statistical analysis. Procedia Eng 64:1602–1611
5. Huang HC, Hsu HS (2015) Biologically inspired deoxyribonucleic acid soft computing for inverse kinematics solver of five-DOF robotic manipulators. Soft Comput 19(4):875–881
6. Soylak M, Oktay T, Turkmen I (2017) A simulation-based method using artificial neural networks for solving the inverse kinematic problem of articulated robots. Proc Inst Mech Eng Part E J Process Mech Eng 231(3):470–479
7. Ayyıldız M, Çetinkaya K (2016) Comparison of four different heuristic optimization algorithms for the inverse kinematics solution of a real 4-DOF serial robot manipulator. Neural Comput Appl 27(4):825–836
8. Cai ZF, Huang DP, Liu YQ (2016) Inverse Kinematics of Multi-joint Robot Based on Particle Swarm Optimization Algorithm. J. Autom. Control Eng. 4(4):305–308
9. Shi Q, Xie J (2018) A research on inverse kinematics solution of 6-DOF robot with offset-wrist based on Adaboost neural network. In: IEEE Int. Conf. Cybern. Intell. Syst. CIS 2017 IEEE Conf. Robot. Autom. Mechatronics-Proc. pp 370–375
10. Zhou Z, Guo H, Wang Y, Zhu Z, Wu J, Liu X (2018) Inverse kinematics solution for robotic manipulator based on extreme learning machine and sequential mutation genetic algorithm. Int J Adv Robot Syst 15(4):1–15
11. El-Sherbiny A, Elhosseini MA, Haikal AY (2018) A comparative study of soft computing methods to solve inverse kinematics problem. Ain Shams Eng J 9(4):2535–2548
12. Dereli S, Köker R (2020) Calculation of the inverse kinematics solution of the 7-DOF redundant robot manipulator by the firefly algorithm and statistical analysis of the results in terms of speed and accuracy. Inverse Probl Sci Eng 28(5):601–613
13. Zhang L, Xiao N (2019) A novel artificial bee colony algorithm for inverse kinematics calculation of 7-DOF serial manipulators. Soft Comput 23(10):3269–3277
14. Shastri S, Parvez Y, Chauhan NR (2020) Inverse kinematics for a 3-R robot using artificial neural network and modified particle swarm optimization. J Inst Eng Ser C 101(2):355–363
15. Deng H, Xie C (2021) An improved particle swarm optimization algorithm for inverse kinematics solution of multi-DOF serial robotic manipulators. Soft Comput 25(21):13695–13708

16. Yiyang L, Jiali X, Hongfei B, Zhining W, Liangliang S (2021) A general robot inverse kinematics solution method based on improved PSO algorithm. IEEE Access. 9:32341–32350
17. Rahkar Farshi T (2021) Battle royale optimization algorithm. Neural Comput Appl 33(4):1139–1157
18. Šegota SB, Anđelić N, Mrzljak V, Lorencin I, Kuric I, Car Z (2021) Utilization of multilayer perceptron for determining the inverse kinematics of an industrial robotic manipulator. Int J Adv Robot Syst 18(4):1–11
19. Mirjalili S (2015) Moth-flame optimization algorithm: A novel nature-inspired heuristic paradigm. Knowledge-Based Syst 89:228–249
20. Bahrami M, Bozorg-Haddad O, Chu X (2018) Moth-flame optimization (MFO) algorithm. Stud Comput Intell 720:131–141
21. Jangid MK, Kumar S, Singh J (2021) Trajectory tracking optimization and control of a three-link robotic manipulator for application in casting. Int J Adv Technol Eng Explor 8(83):1255–1267
22. Jangid MK, Kumar S, Singh J (2022) Inverse kinematics solutions of a newly designed three-link robotic manipulator for the casting process using ant lion optimizer. Int J Adv Technol Eng Explor 9(97):1704–1717

Development of Solar-Powered Electric Vehicle Prototype

Suvarna Rode, Rashmi Kale, Devoushka Yadav, Dhananjay Pawar, Jatin Satyam, and Nishant Shelar

Abstract The subsequent use of fossil fuels for automobiles has increased the emission rates, deteriorating the quality of the environment. Increased use of automobiles has caused scarcity of fossil fuels hence giving rise to usage of renewable energy sources. In order to generate electricity to charge electric vehicles, the power has to still be extracted from fossil fuels hence to provide abundant power source, renewable energy is used. This paper focuses on the usage of solar power to charge a prototype of an electric vehicle. Since solar power independently cannot be directed to the power system of the vehicle, a lead acid battery with a charge controller was incorporated into the vehicle in order to store and charge the system.

Nomenclature

3D CAD	Three-dimensional computer-aided design
DOD	Depth of discharge
EV	Electic vehicle
SPEV	Solar-powered electric vehicle

S. Rode · D. Yadav (✉) · D. Pawar · J. Satyam · N. Shelar
Mechanical Engineering Department, Fr. C. Rodrigues Institute of Technology Vashi, Navi Mumabi, Maharashtra 400703, India
e-mail: devoushka.Yadav@gmail.com

R. Kale
Electrical Engineering Department, Fr. C. Rodrigues Institute of Technology Vashi, Navi Mumabi, Maharashtra 400703, India

A. Sachdeva et al. (eds.), *Recent Advances in Operations Management and Optimization*, Lecture Notes in Mechanical Engineering,
https://doi.org/10.1007/978-981-99-7445-0_16

1 Introduction

Electric vehicles have become more prevalent in the automobile industry with the recent advancements and need of technology. However, they are constrained by some limitations that restrict their growth in the market entirely. The major issue that comes across is of charging and its power storing capacity (battery) [1, 2]. Charging EVs with grid power utilizes energy extraction by conventional methods which is harnessing it from fossil fuels or power plants. The electricity that comes from the grid is unclean, inefficient, and needs more time to charge an EV [4, 7]. Charging stations need more space, time, and high power supply in order to charge EVs. The daily human needs have expanded up to a large extent where transportation serves as the basic amenity. Excess use of private vehicles has caused an immense increase in fuel usage and pollution rate. This high rate of pollution has changed various environmental factors and increased global warming rates making it hazardous for the future. Excess usage or burning of fossil fuels alters geographic patterns majorly leading to acid rain formation hence destroying terrestrial areas of crop generation [5, 6]. These factors have indirectly caused an economic imbalance as well which causes investing in repairing more rather than creating. Not only has the use of fossil fuels caused environmental problems but the methods of extraction have also led to various alterations. The extraction of crude oil from the oceans causes severe and fatal impacts on the marine life. Cyclic extractions cause uneven bed rocks and lead to pollution of groundwater level. There are serious risks of oil spills in the ocean while transporting or extracting fuel which is again harmful.

To overcome this problem renewable sources of energy are found to be a sound alternative to present and future energy crises [3, 8]. Solar energy exerts energy from the sun in the form of solar diffusion for heat or to produce electricity [1].

This paper focuses on the integration of renewable energy with automobiles. The objective of the paper is to design a prototype of a solar vehicle where the energy to charge the vehicle is purely extracted from solar panels. Since solar energy is the most clean and abundant source of renewable energy and its panels are flexible in application these can be easily mounted on the roofs and bonnets of vehicles. In this work, we implanted the solar panel on the roof of the prototype. Here the solar energy is converted into electrical energy by using photovoltaic cells and is further sent to the battery which stores the charge for its further use. This cyclic charging is regulated with the help of a charge controller which monitors the flow of energy into the battery in order to maintain its life and prevent its exposure to fluctuating energy directly from the solar panels. The energy from the battery is then sent to the motors connected to the front wheels of the prototype.

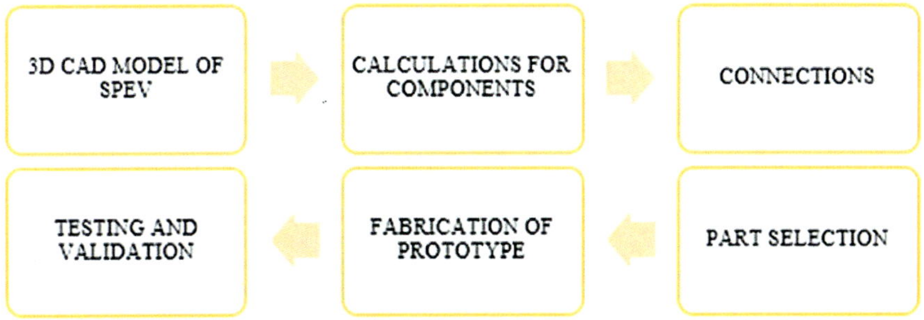

Fig. 1 Flowchart for fabrication of SPEV prototype

2 Methodology

Figure 1 illustrates the working methodology for the prototype fabrication of solar-powered electric vehicles. It is the set of guiding principles and processes involved in the execution of the project. Involves the collection of methods, practices, processes, techniques, procedures, and rules.

Methodology begins with the design of a 3D CAD model of the prototype which had been accomplished after various iterations using Autodesk Inventor software. After completing the design of the model, the components used in the design have been selected according to the calculations performed. These components were then virtually connected and have been shown by the graphical representation. Part selection involves purchasing the components according to the desired values obtained from the calculations. Further, the model was assembled and the prototype was fabricated. This prototype was then tested and validated to our objective.

3 CAD Model, Connections and Calculations

CAD model is a virtual representation of the real-life model. It allows us to build realistic computer models and assemblies allowing us to refine and manipulate the design before its production. Connections or the circuit layout of the system represent the graphical or the visual electrical connections of the product. In order to proceed with the actual fabrication the component with the desired characteristics should be known. Hence, this section of the paper represents the iterative CAD model of the prototype along with the connections of the prototype and the calculations required for the component selection.

Figure 2, represents the final iterative 3D CAD model of the prototype which had been designed in Autodesk Inventor software. It indicates the components required by the SPEV prototype in order to function completely. Figure 3 illustrates the circuit diagram of the vehicle. The connection has been made such that the solar panel's photovoltaic cells convert solar energy into electrical energy, which is then transferred

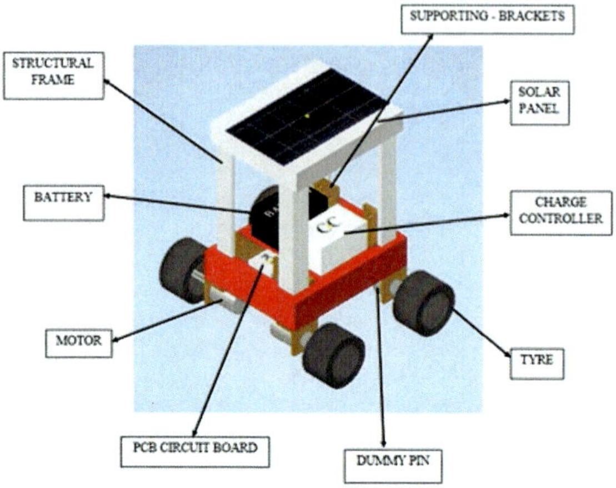

Fig. 2. 3D CAD model of the prototype

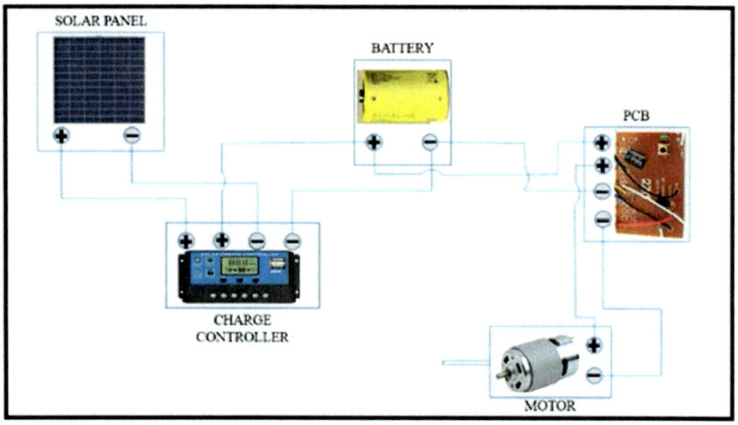

Fig. 3 Circuit layout of the prototype

to the charge controller known as the flow regulator to prevent overcharging or discharging of the battery. It is then provided to the PCB which subsequently transmits signals to the motor, which in turn drives the wheels.

3.1 Torque Calculation

Motor torque is the output of rotational work done by the motor and the SI unit is given as Nm.

Let w be the work done by the motor then

$w = \text{force } (f) * \text{distance moved per revolution } (s)$

$$\frac{w}{s} = \frac{f \times 2\pi r \times N}{60}$$

where N is the number of turns per second and since the electrical power is equivalent to mechanical power. Therefore,

$$W = E \times I = \frac{2\pi N}{60} \times (f \times r)$$

Hence, motor torque is given as

$$\tau = 9.549 \times \frac{\text{power}}{\text{speed}}$$

$$\tau = \frac{9.549 \times 12 \times 0.5\text{Nm}}{6400} = 0.19\text{Nm}$$

3.2 *Solar Panel Calculation*

Load produced by the battery: Battery specifications: lead acid rechargeable

12 V; 1300 mAh (charge capacity)

Power capacity or the load towards solar panel

$$\begin{aligned} &= \text{V} * \text{I} \\ &= 12 * 1.3 \\ &= 15.6 \text{ Wh} \end{aligned}$$

Assuming we get 2 h of direct sunlight,
And hence the solar panel required by the battery to charge itself is

$$\begin{aligned} &= 15.6\,\text{Wh}/2\,\text{h} \\ &= 5.2\,\text{W} \approx 5\,\text{W} \end{aligned}$$

3.3 Battery Charge Time

- Estimating battery charge current: 5 W/12 V = 0.42 A.
- Multiplying battery capacity by rule of thumb losses 0.42*0.8 = 1.08 A.
- Estimated time for battery charging 1.3/0.85 = 1.53 Ah.
- Battery charge time 1.53/0.42 = 3.64 h.
- Depth of discharge of battery 3.64*0.5 = 1.82.
- Charge absorption time 1.82 + 2 = 3.82 h.

Hence, the approximate time taken by the battery to charge is 4 h.

3.4 Battery Discharge Time

It is given as the ratio of battery capacity to the current consumption of the load.

Ah generated by battery/A consumed by load.

Battery capacity = 1.3 Ah.

Consumed by load = 0.5 A.

$$1.3\,\text{Ah}/0.5\,\text{A} = 0.26\,\text{h}$$

Hence the total discharge time of the battery is 15 min.

4 Fabrication

Figures 4, 5 and 6 represent the top view, side view and top view of the project, respectively. It is the practical implementation and integration of all the components with the interpretation of results. The prototype has been designed at its lowest height such that its center of gravity is closer to its geometric center and the weight has been adjusted to avoid tipping of vehicle.

Components of the prototype:

1. Solar panel: a system of multiple photovoltaic cells made up of photosensitive material that emits energy in the presence of sunlight further transmitted to electrons in order to overcome potential barriers.
2. Charge controller: is an energy monitor that maintains and controls the overcharging and discharging of the battery
3. Battery: stores the energy received from the solar panel
4. Motor controller: regulates the input energy that is transmitted to the motors
5. Motor: DC gear motors provide sped control of the vehicle as it can deliver high torque at low speed as the gearhead functions as a torque multiplier.

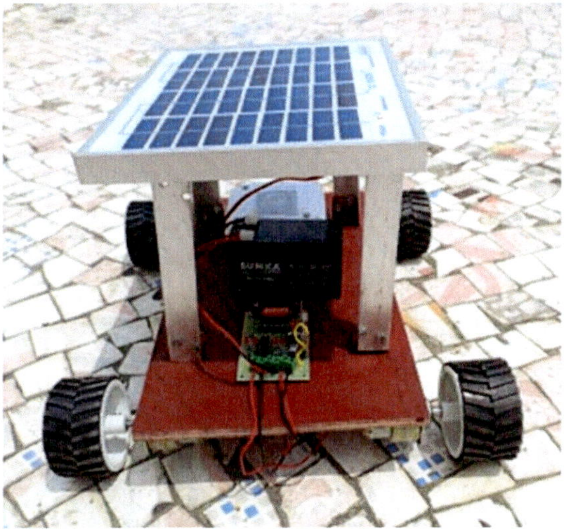

Fig. 4 Front view of the prototype

Fig. 5 Side view of the prototype

5 Results and Discussions

Battery charging depends majorly on the chemical composition of the battery. The battery that had been used was a 12 V lead acid battery which allows 50% DOD that is the battery should not be discharged or used more than 50% of battery capacity in order to avoid self-life of the battery. For a 12 V lead acid battery to be 50% discharged the voltage should be 12.05 V and for complete discharge that is at 10% DOD the minimum voltage should be 10.5 V [9]. Our prototype has been tested for different DODs which is shown below.

Table 1, indicates the charging time taken by the battery to reach 100% charging from a certain state of charge. It illustrates that for the battery to charge 40% it took 13 min whereas to charge 60% it took 15 min.

Fig. 6 Top view of the prototype

Table 1 Charging time required by battery

State of charge (%)	Minimum voltage (before charging)	Maximum voltage (after charging)	Charging time (in min)
60	12.2	12.7	13
40	11.9	12.6	15

Table 2 shows the discharge time taken by the prototype vehicle in order to discharge at different states of charge. For 40% discharge of the battery capacity it took 11.3 min and for 60% discharge it took 18.8 min.

From the above data, the following can be inferred

1. Average charging time of the battery: 14 min.
2. Average discharging time of the battery: 15.05 min.

Table 2 Discharging rate of the battery

State of charge (%)	Voltage level	Discharging time (in min)
60	12.5–12.2	11.3
40	12.72–12.03	18.8

6 Conclusions

Solar-powered charging is a cleaner and convenient option for charging as it is a self-adaptive method of charging. Integrated photovoltaic cells make it convenient for the system to absorb renewable energy and convert it to electrical energy, which in turn charges the engine of the prototype vehicle. This type of charging is advantageous over grid charging systems as it allows faster charging as compared to conventional charging systems. The abundant solar energy makes this system more suitable as it is portable and easier in terms of application. This paper proposes the impact of renewable charging in automobiles as it provides unadulterated and fast-charging fuel to its engine. The main motive of SPEV is to portably charge its system by using a renewable energy source, i.e., solar power. The prototype fabrication begins with generating a 3D CAD model of the prototype after various iterations, to drive the vehicle it has to overcome the moment of inertia and other resisting forces hence suitable torque calculated is 0.19 Nm. To drive this torque external electromotive force of a 12V lead acid battery is applied which is further charged by a solar panel of 5 watts. The components are selected based on the calculations and the prototype fabrication was completed. It was further tested to its objectives and the results were analyzed. After testing and validation, the average charging time of lead acid battery came out to be 14 min with discharging time of 15.05 min.

References

1. 1st international conference on electrical engineering and information & communication technology (ICEEICT 2014): military institute of science and technology (MIST): 10–12 April 2014 Military Institute of Science and Technology. Department of Electrical E, Jāhāṅgīranagarabiśvabidyālaẏa. Institute of Life Sciences Information Technology, Institute of Electrical and Electronics Engineers
2. Chaudhari K, Ukil A, Kollimalla S (2018) Hybrid optimization for economic deployment of ESS in PV-integrated EV charging stations. IEEE Trans Ind Inform 14(1):106–116
3. Akindele JIMBA J, Simon A, Egbe E (2018) Development of an indigenous solar powered car. Int J Sci Eng Res 9(11)
4. RVS Technical Campus (2019) Proceedings of the third international conference on electronics, communication and aerospace technology (ICECA 2019), 12–14, June 2019. IEEE Aerospace and Electronic Systems Society, Institute of Electrical and Electronics Engineers
5. Kundu A (2020) IEEE international conference for convergence in engineering: proceedings, 5–6 September, Kolkata, India. Institute of Electrical and Electronics Engineers, Kolkata Section
6. Singh B, Verma A, Al-Haddad K (2018) Implementation of solar PV-battery and diesel generator based electric vehicle charging station. In: Proceedings of 2018 IEEE international conference on power electronics, drives and energy systems, PEDES
7. Solar EV Charging Shatnawi M, Ari K, Alamoodi O (2021) 2021 6th international conference on renewable energy: generation and applications, ICREGA
8. Zhao H (2014) An intelligent solar powered battery buffered EV charging station with solar electricity forecasting and EV charging load projection functions. In: Burke A2014 IEEE international electric vehicle conference, IEVC
9. Larminie J, Lowry J (2012) Electric vehicle technology explained. Wiley, p 342

Recent Biotechnological Approaches for Plastic Waste Management

Amit Dhaundiyal, Virangna Jamwal, and Aanchal Mittal

Abstract Biotechnological approaches have emerged as an efficient and promising solution for the management of PW across the globe. The annual production of plastics has drastically increased in the recent few years. Numerous methods are employed for the disposal of PWs such as recycling, incineration, and landfilling. Even though recycling is a good waste management strategy it also has its own drawbacks. During recycling, many toxic pollutants are released directly into the environment, which are very harmful to the whole ecosystem. Additionally, incineration accounts for the release of toxic gases in the environment and landfills require a huge proportion of land which is not a great attribute for proper waste management. Here biotechnology comes into play as it is a more suitable and sustainable approach for proper waste management of plastic polymers without harming the natural ecosystem. These approaches utilize microorganisms such as bacteria, actinomycetes, and fungi to degrade the plastic polymers into monomers which are further used in the recycling process. These microorganisms have the ability to break down complex polymers into simpler molecules which can be used as a source of energy or as a raw material for the production of new plastics. The most commonly used method for PW management is biodegradation where the plastic polymers are degraded by the action of microbes into simpler molecules thus eliminating the PW from the environment. Biodegradation of plastics by microbes cannot occur or occurs at a very slow rate if the prior degradation by agents like UV radiation, water abrasion, photooxidation, and corrosion is not subjected. Additionally, enzymatic degradation is also commonly utilized nowadays where the enzymes extracted from different microbial sources degrade the synthetic polymers which aids in the removal of pollutants from the environment. Using biotechnology for the modification of these

A. Dhaundiyal · V. Jamwal
Research Scholar, Department of Bioscience and Technology, Vellore Institute of Technology, Vellore, Tamil Nadu 632014, India

A. Mittal (✉)
Assistant Professor, Department of Bioscience and Technology, Vellore Institute of Technology, Vellore, Tamil Nadu 632014, India
e-mail: aanchalmittal0411@gmail.com

A. Sachdeva et al. (eds.), *Recent Advances in Operations Management and Optimization*, Lecture Notes in Mechanical Engineering,
https://doi.org/10.1007/978-981-99-7445-0_17

enzymes and enhancing their degradational efficiency is the key approach to overcoming the PW problem throughout the world. This review highlights the different biotechnological approaches for the efficient and sustainable degradation of plastic polymers thus contributing to PW management efficiently around the world.

Nomenclature

PW	Plastic waste
PE	Polyethylene
PET	Polyethylene terephthalate
PUR	Polyurethane

1 Introduction

Plastics are a widely used commodity due to their durability, lightweight, cost-effectiveness, and inertness as well as good thermal and electrical insulation [1]. The ever-increasing use of plastic products in different sectors has generated a huge amount of waste throughout the world. In 2015, approximately 8300 million metric tonnes (Mt) of PW had been generated out of which only 9% had been recycled, 12% was incinerated, and the rest 79% was accumulated in the natural ecosystems [2]. PW management has become a major global challenge due to the increasing production and consumption of plastic products, leading to a large amount of PW that is difficult to degrade and poses a huge threat to the environment and living organisms. In the environment, plastic degradation occurs due to the action of UV radiation, wind abrasion, photooxidation, thermal degradation, and the action of waves which results in the alteration of their physical and chemical properties making them weaker, brittle, and more susceptible for microbial degradation.

The conventional methods of PW management currently in use are recycling, incineration, and landfilling; however, these methods are not effective as they lead to the leaching of hazardous pollutants and toxic gases into the environment. During recycling, the melting of plastic releases sulfur, carbon dioxide, and other gases into the environment which accounts for acid rain, global warming, greenhouse effect [3]. Thus, there is a need for sustainable waste management for the efficient degradation of plastic polymers throughout the world. Biodegradation involves the action of microorganisms such as bacteria, fungi, and actinomycetes for the degradation of plastic polymers. In recent years several microbes and enzymes have been identified that are able to degrade synthetic plastics. The microbes are acknowledged for their survivability and can thrive in different environmental conditions owing to these varied habitats they possess a variety of catabolic pathways which might prove to be of use in degrading polymers in accordance with their properties [4].

Several classes of enzymes are known for the degradation of long chains of carbon-hydrogen polymers into oligomers and monomers. These enzymes adhere to the plastic surface through hydrophobic interactions and then break them into small monomers which are then utilized by these microbes for their growth and metabolism. In recent years the extraction and modification of enzymes have proved to be a promising solution for the degradation of synthetic plastics by the usage of different techniques.

The main scope of this work is to summarize the conventional practices of PW disposal and the microbes responsible for the breakdown of synthetic plastic polymers. In addition to these various biotechnological approaches for the waste disposal system, micro-organism-assisted degradation and enzymes that are potent for degrading plastics are also discussed.

2 Current Strategies for PW Disposal

PW is generated in huge amounts due to the growth of various sectors of industries and human population which thereby causes huge demand for plastics for various purposes in different sectors, which include industries, household, agriculture, medical, etc. (shown in Fig. 1). Government authorities have established several policies and regulations for the proper disposal of plastic materials. This includes measures such as endorsing recycling, reducing single-use plastics, improving the waste management system, and imposing fines for littering and illegal dumping. Some conventional methods for PW disposal are recycling, incineration, and landfilling. Recycling of PWs includes collecting, sorting, cleaning, and reprocessing PW for producing new products. It is an important part of PW management as it helps in reducing the amount of PW and also it accounts for less emission of CO_2 and other gases which are produced during the incineration or combustion of plastics [5]. However, chemicals released in recycling areas have a negative impact on plants and animals living nearby. Incineration refers to the combustion of PWs at high temperatures around 450–700 °C in the presence of oxygen for the generation of energy. Plastic wastes are very similar to hydrocarbon-based fuels thus they yield a good amount of energy however during this process carbon dioxide is released into the atmosphere which causes a serious climatic crisis [6]. Additionally, the burning of PWs also releases certain toxic gases, heavy metals, and ash waste into the environment which can be bad for health [7]. The unburnt plastic remains in the form of microplastics, which leach out and release to a different ecosystem. PW after being used ends up in landfills, where the plastics are dumped in the earth's face which affects the land due to its accumulation [8]. The waste is compacted and layered to minimize the amount of space it takes and to prevent the spread of odors and diseases. PW in landfills can take hundreds of years to decompose and release harmful chemicals into the environment, contaminating soil and groundwater [9]. Additionally, landfills often generate methane, a potent greenhouse gas as a result of the breakdown of organic matter.

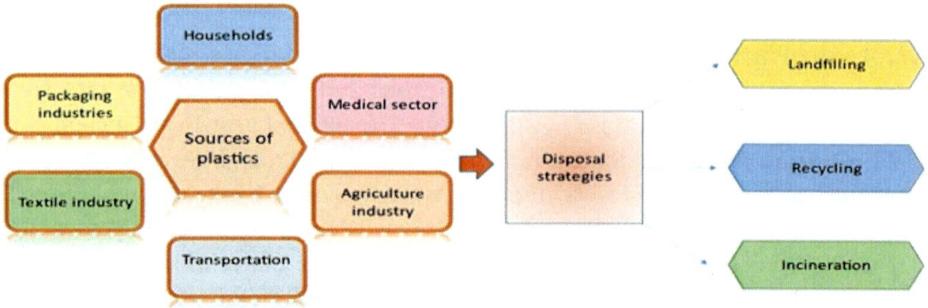

Fig. 1 Sources and disposal of plastic waste

3 Degradation of Plastic by Microorganisms

There are several microorganisms that are known for their potent degradation efficiency of hydrocarbon-based polymers and pollutants that end up as waste in the environment. Most plastic materials are hydrophobic in nature thus the microbial attachment to the surface is the primary step for their degradation. They form biofilms around the surface of plastic polymers thereby facilitating biodegradation. The overall process of biodegradation of plastics is divided into four stages [10]. Biodeterioration is the process where plastic polymers are subjected to changes in their physical, chemical, and mechanical properties by forming biofilms around the surfaces. Bio-fragmentation involves the degradation of plastic polymers by the enzymatic action of microbes which results in hydrolysis of bonds present in the polymers. Assimilation accounts for the uptake of fragmented molecules into the microbial cells through biomembranes where the monomers are oxidized and energy is produced which is utilized by the microbes for their growth. Mineralization is the final step in this process where the atoms are incorporated within the microbial cells and the oxidized products are released in the form of CO_2, N_2, H_2O, and CH_4 [11].

Several microbes have been isolated and reported for the efficient degradation of plastic polymers. Some commonly distributed species of Pseudomonas sp., bacillus sp., and *Ideonella sakaiensis* sp. have been reported for the degradation of synthetic plastics (Table 1). These strains were commonly isolated from the municipal dump sites contaminated with plastic waste and found to be potent for their degradation. The degradation of PET is relatively challenging because of its complicated nature.

As shown in the table out of all microorganisms, the degradation of PET was shown maximum by *Ideonella sakaiensis* sp. which was nearly about 75% [12]. Consortia of *Sarcina aurantiaca, Bacillus subtilis, Aspergillus flavus*, and *Aspergillus niger* also degrades PET significantly, which was around 28.78% weight reduction in 60 days of incubation [13]. Similarly, the degradation of PE was considerably shown by ascomycetes [14]. Several strains of Bacillus sp. were reported for the degradation of PE films from the dumping sites [15, 16].

Table 1 Microorganisms involved in plastic degradation

Microorganism involved	Substrate	Observed degradation (%)	Incubation duration	Source	References
Consortia of *Sarcina aurantiaca, Bacillus subtilis, Aspergillus flavus and Aspergillus niger*	PET	Observed 28.78% weight loss	60 days	Kauvery River, Srirangam (South India)	[13]
Ideonella sakaiensis	PET	75% degradation of PET films	NR	PET-contaminated dumping site	[12]
Achromobacter x ylosoxidans	HDPE Film	9% weight reduction	50 days	Landfill site soil	[17]
Enterobacter and Pseudomonas	LDPE	12.5 and 15%	150 days	Plastic dumping site Karnataka, India	[18]
Pseudomonas aeruginosa strain ISJ14	PE	Observed 8.7% weight reduction	60 days	NR	[19]
Aspergillus terreus, Aspergillus sydowii (Ascomycete)	PE	Observed 28.4% weight reduction	NR	NR	[16]
Acinetobacter baumannii	PET	27.3% weight reduction	90 days	Soil and PW sample	[20]
Bacillus velezensis Bacillus subtilis	PE	Observed 8.01%, 9.26% respectively	90° days, 30° days respectively	Dumping sites	[15, 16]
Pseudomonas sp.	LDPE	Observed 5% weight reduction	45 days	Municipal dumpsite	[21]

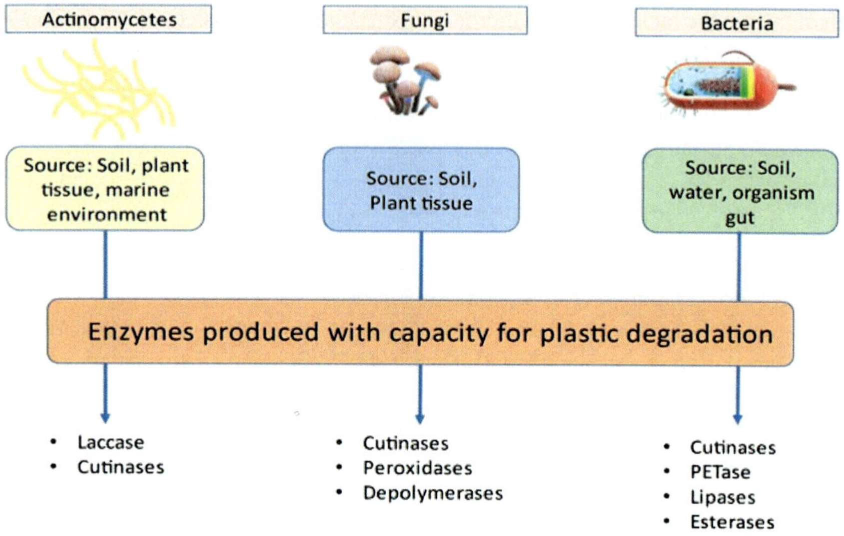

Fig. 2 Microbes and their enzymes used for plastic degradation

4 Enzymatic Degradation of Plastic

The use of enzymes in the degradation of plastic is considered a promising strategy for the sustainable method of plastic waste management. There are several classes of enzymes that are known to be capable of breaking plastics including lipases, cellulases, proteases, esterases, cutinases, etc. (Fig. 2).

These enzymes hydrolyze the polymer chains present in plastics, breaking them into oligomers and monomers that can be further degraded by microbes (Table 2). PETase and MHETase enzymes isolated from *Ideonella sakaiensis* act synergistically on PET films lead to their degradation into smaller monomeric units [22]. Similarly, pseudomonas sp. has been reported to degrade plastic by utilizing different enzymes. *Pseudomonas aeruginosa* strain DSM 50071 was reported to secrete serine hydrolases which are responsible for the breakdown of polystyrene [23]. PUR was degraded by the enzyme (Lipase) secreted by *Pseudomonas protegens* cultured in lysogeny broth-Miller (LB) supplemented with glucose [24].

Table 2 Microbial enzymes for plastic degradation

Microbes	Plastic for degradation	Enzyme extracted	Method of detection	References
Ideonella sakaiensis	PET	MHETase and PETase	Assessment of degradation product	[22]
Pseudomonas aeruginosa DSM 50071	PS	Serine hydrolase	FTIR, NMR	[23]
Cryptococcus sp.	PBS	Lipase	FTIR, NMR	[25]
Humicola insolens and Thermobifida cellulosilytica	PBTF and PBF	Cutinase	Percentage weight loss with SEM	[26]
Pseudomonas protegens	PUR	Lipase	NMR, Zymography	[24]
Fusarium solani	PBS	Cutinase	XRD SEM	[27]
Aspergillus fumigatus	PHB, PBS, PES	Polyhydroxybutyrate depolymerase	Zymography	[28]

5 Conclusion

Biotechnological approaches in plastic waste management hold great promise for addressing the plastic waste crisis and mitigating its environmental impacts. These approaches utilize microbes and enzymes that break down plastic into smaller biodegradable compounds. This not only reduces the amount of plastic waste in the environment but also provides an alternative to conventional methods of plastic disposal such as incineration and landfilling, which have negative environmental impacts. One of the major benefits of biotechnological approaches is to break complex plastics such as PE and PP, which are highly resistant to degradation by other means. These approaches can also be more cost-effective and sustainable as compared to conventional methods, as they do not require high energy inputs or produce toxic by-products.

References

1. Liang Y, Tan Q, Song Q, Li J (2021) An analysis of the plastic waste trade and management in Asia. Waste Manag 119:242–253
2. Geyer R, Jambeck JR, Law KL (2017) Production, use, and fate of all plastics ever made. Sci Adv 3(7):e1700782
3. Shen L, Worrell E (2014) Plastic recycling. In: Handbook of recycling. Elsevier, pp 179–190
4. Moharir RV, Kumar S (2019) Challenges associated with plastic waste disposal and allied microbial routes for its effective degradation: A comprehensive review. J Clean Prod 208:65–76

5. Vollmer I, Jenks MJ, Roelands MC, White RJ, van Harmelen T, de Wild P, ... , Weckhuysen BM (2020) Beyond mechanical recycling: Giving new life to plastic waste. Angew Chem Int Ed 59(36):15402–15423
6. Shome R (2020) Role of microbial enzymes in Bioremediation. ELifePress 1:15–20
7. Netzer C, Li T, Løvås T (2021) Surrogate reaction mechanism for waste incineration and pollutant formation. Energy Fuels 35(9):7030–7049
8. Payne J, McKeown P, Jones MD (2019) A circular economy approach to plastic waste. Polym Degrad Stab 165:170–181
9. Thiounn T, Smith RC (2020) Advances and approaches for chemical recycling of plastic waste. J Polym Sci 58(10):1347–1364
10. Montazer Z, Habibi Najafi MB, Levin DB (2020) Challenges with verifying microbial degradation of polyethylene. Polymers 12(1):123
11. Amobonye A, Bhagwat P, Singh S, Pillai S (2021) Plastic biodegradation: frontline microbes and their enzymes. Sci Total Environ 759:143536
12. Yoshida S, Hiraga K, Takehana T, Taniguchi I, Yamaji H, Maeda Y, ... , Oda K (2016) A bacterium that degrades and assimilates poly(ethylene terephthalate). Science 351(6278):1196–1199
13. Maheswaran B, Al-Ansari M, Al-Humaid L, Raj JS, Kim W, Karmegam N, Rafi KM (2023) In vivo degradation of polyethylene terephthalate using microbial isolates from plastic polluted environment. Chemosphere 310:136757
14. Ghosh SK, Pal S, Ray S (2013) Study of microbes having potentiality for biodegradation of plastics. Environ Sci Pollut Res 20:43394355
15. Nehal N, Singh P (2021) Optimization of cultural condition of Bacillus sp. MZ540316: improve biodegradation efficiency of lipopeptide biosurfactant against polyethylene. Biomass Convers Biorefinery: 1–17
16. Liu X, Zhang Y, Sun Q, Liu Z, Zhao Y, Fan A, Su H (2022) Rapid colonization and biodegradation of untreated commercial polyethylene wrap by a new strain of Bacillus velezensis C5. J Environ Manage 301:113848
17. Kowalczyk A, Chyc M, Ryszka P, Latowski D (2016) Achromobacter xylosoxidans as a new microorganism strain colonizing high-density polyethylene as a key step to its biodegradation. Environ Sci Pollut Res 23:11349–11356
18. Skariyachan S, Taskeen N, Kishore AP, Krishna BV, Naidu G (2021) Novel consortia of enterobacter and pseudomonas formulated from cow dung exhibited enhanced biodegradation of polyethylene and polypropylene. J Environ Manage 284:112030
19. Gupta KK, Devi D (2020) Characteristics investigation on biofilm formation and biodegradation activities of *Pseudomonas aeruginosa* strain ISJ14 colonizing low density polyethylene (LDPE) surface. Heliyon: 6(7):e04398
20. Hussein AA, Alzuhairi M, Aljanabi NH (2018) Degradation and depolymerization of plastic waste by local bacterial isolates and bubble column reactor. In: AIP conference proceedings, vol 1968, no 1. AIP Publishing LLC, p 030081
21. Tareen A, Saeed S, Iqbal A, Batool R, Jamil N (2022) Biodeterioration of microplastics: A promising step towards plastics waste management. Polymers 14(11):2275
22. Knott BC, Erickson E, Allen MD, Gado JE, Graham R, Kearns FL, ... , McGeehan JE (2020) Characterization and engineering of a two-enzyme system for plastics depolymerization. Proc Natl Acad Sci 117(41):2547625485
23. Kim HR, Lee HM, Yu HC, Jeon E, Lee S, Li J, Kim DH (2020) Biodegradation of polystyrene by Pseudomonas sp. isolated from the gut of superworms (larvae of Zophobas atratus). Environ Sci & Technol 54(11):6987–6996
24. Hung CS, Zingarelli S, Nadeau LJ, Biffinger JC, Drake CA, Crouch AL, ... , Crookes-Goodson WJ (2016) Carbon catabolite repression and Impranil polyurethane degradation in Pseudomonas protegens strain Pf-5. Appl Environ Microbiol 82(20):6080–6090
25. Thirunavukarasu K, Purushothaman S, Sridevi J, Aarthy M, Gowthaman MK, Nakajima-Kambe T, Kamini NR () Degradation of poly (butylene succinate) and poly (butylene succinate-co-butylene adipate) by a lipase from yeast Cryptococcus sp. grown on agro-industrial residues. Int Biodeterior Biodegrad: *110*, 99–107 (2016).

26. Gigli M, Quartinello F, Soccio M, Pellis A, Lotti N, Guebitz GM, ... , Munari A (2019) Enzymatic hydrolysis of poly (1, 4-butylene 2, 5-thiophenedicarboxylate) (PBTF) and poly (1, 4-butylene 2, 5-furandicarboxylate) (PBF) films: a comparison of mechanisms. Environ Int 130:104852
27. Hu X, Gao Z, Wang Z, Su T, Yang L, Li P (2016) Enzymatic degradation of poly (butylene succinate) by cutinase cloned from Fusarium solani. Polym Degrad Stab 134:211–219
28. Jung HW, Yang MK, Su RC (2018) Purification, characterization, and gene cloning of an Aspergillus fumigatus polyhydroxybutyrate depolymerase used for degradation of polyhydroxybutyrate, polyethylene succinate, and polybutylene succinate. Polym Degrad Stab 154:186–194

Reviewing Enablers and Drivers While Implementing Artificial Intelligence (AI) Among Indian Automobile Supply Chains

Eisha Mehta Sharma and Bikram Jit Singh

Abstract Since the economic liberalization of the Indian market in the early 1990s, the Indian automobile industry has grown abruptly which consists of vehicle and component manufacturers. The global auto companies have urged the domestic sector to adopt supply chain practices. The advancement in technology has made it possible for the organization to expand its business globally without the addition of any surplus labor, space, area, etc. The decision support, automation, predictive and prescriptive outcomes are well transformed by Artificial Intelligence in supply chains. AI has been proven as a fortune in the world of automation for industrial setups. Without its help, tasks such as smart decision-making, research and data analysis, minimizing errors and business continuity would not have been trouble free. The Indian Automobile supply chains have been broadly classified into three categories namely; Passenger Vehicles, Commercial Vehicles and Two/Three Wheeler Vehicles respectively. This research paper aims to find the enablers and drivers that boost up the path of implementation of AI before the Raw material suppliers; component manufacturers-sub assembles, distribution channels (comprising of dealers, retailers, service providers and finally the end users (customers). The enablers and their effects are deeply analyzed for all the three above-stated segments of the Indian automobile supply chains.

Keywords Supply chains system · Artificial Intelligence · Enablers · Drivers · Success factors · Automobile sector · Manufacturing chains · Commercial vehicles · Passenger vehicles · Two/three wheeler vehicles

E. M. Sharma (✉) · B. J. Singh
Department of Mechanical Engineering, MMEC, Maharishi Markandeshwar Deemed to be University, Mullana, India
e-mail: eishamehta1208@gmail.com

A. Sachdeva et al. (eds.), *Recent Advances in Operations Management and Optimization*, Lecture Notes in Mechanical Engineering,
https://doi.org/10.1007/978-981-99-7445-0_18

1 Introduction

From the early years of the 1990, during the time of economic liberalization, the Indian automobile industry has grown very speedily. The introductions of major global automobile industries have enhanced the usage of supply chain systems in the best possible ways. A huge increase in exports is witnessed which has boosted the competitiveness in the market. However, the Indian automobile industry has to work in a quirky environment thereby increasing the challenges to the prevailing complex automobile supply chains [1].

The automobile industry is definitely the widest ranging trade in the international economy and has been proven as a shining sector for the Indian economy. The Indian economy has placed a prominent position in the automotive sector [2]. In the field of manufacturing, it has been proven as the leader in process, product and technology. A nation's economic health is proved by the performance of its automotive sector and its growth [3]. Presently the automobile sector is facing several challenges and provocations which include the urge to reduce the cost due to the customer's expectations, production of international products at local prices, flawless performance of final product without any fault in working to keep away from innumerable penalties, managing the demands due to fluctuating needs, unification with the customers at new product development phase, problems in revealing the private information about the new designed product, visibility and accuracy of the supply chain, in terms of inventory and documents, flexibility of the supply chains [4].

The prevailing government norms in India and the worldwide slowdown in business have made the automobile industry go through a catastrophe after a long time. A post-effect of COVID-19 [5]. A note has to be made that the automobile industry has always changed the procedures, methods and technology after every global slowdown. Many technological aids are being incorporated by the automobile industry in the present day. Certain software tools like Enterprise Resource Planning, Supply Chain Management, Customer relationship management, and some tangible tools like Universal product code, Global Positioning Systems, interfaces, Electronic Data Interchange, cyberspace, prescribed tools like outsourcing, Vendor Management Inventory, etc. These aid in reducing the cost by modifying the preexisting procedures and recurrent investigations and evaluations like audits and VSMs [6]. To enhance the future demand in the automobile industry there's a constant requirement for cost reduction so to meet the necessity of time new technologies like artificial intelligence, machine learning, data science, cloud computing, smart manufacturing, industrial internet of things and robotics are being implemented in the Indian automobile SCMs [7]. Here in this paper focus is made on the field of AI in Indian Automobile SCM, particularly finding the barriers that come in the path of AI while being implemented in the field of Indian Automobile supply chains [8]. The technology of making programmed machines think like human minds and make decision in the manner the human brain does is termed as artificial intelligence. The main feature of AI is that it can justify and perform actions that fulfill a specific task in best suitable way [9].

The problem-solving and decision-making power of the human brain is attached to the computer by means of AI [10]. According to Alan Turing, to the task of problem-solving AI is the field that amalgamates computer science and vigorous databases. The sub-field of ML and deep learning is also circumscribed. That is generally mentioned in the context of AI. Machine learning and deep learning are made up of AI algorithms. That creates a fine system giving the predicted results. These predicted results are based on input data [11]. Enabling the use of large data sets eradication of manual human intervention is brutalized by ML [12]. In this world of information technology, the vigorous needs of customers can only be satisfied by providing rapid and authentic products or services on time, through effective and efficient Supply Chains, which is not possible without proper AI intervention [13].

2 Literature Survey

Without any exceptions, the most powerful driving growth factor and the major growth contributor in any nation's economy remains the automobile sector. The automobile has been described as 'both form and function' based product because it has the involvement of peer engineering processes along with a presentable appearance [14]. The automobile industry has been termed as the 'industry of industries' due to the involvement and usage of final products of all the manufacturing industries thereby supporting other businesses like mining, steel, etc., i.e., upstream business and fields of insurance, finance, etc., i.e., downstream business [15]. India, Brazil and China are the prominent rising economic powers that have considerable domestic demands and a requisite local production [16]. The Indian automobile industry had its commencement in the 40 s and it has had substantial growth in the past twenty years because of economic modernization with a cent percent FDI in this sector [17]. The automotive industry is widely fragmented into three divisions which are commercial vehicles, passenger vehicles and two or three-wheelers. The automobile supply chain is complicated and compounded because of the involvement of the provider of raw material, component processors, sub-assemblers, final fabricator, various distribution passages, networks and concluding consumers, i.e., customers [18]. All these above-stated stakeholders at different levels contribute importantly to increasing sales and profits for the automobile industry. So there becomes a need for the vigorous and powerful SCM that aims to unite all the above-stated stakeholders to increase cost efficiency and operating profits. In this world empowered by information and technology, no other tool than AI artificial intelligence can be a helping hand in the supply chains for the ease of working and increased productivity[19]. Simchilevi, Kaminsky, and Simchi-levi in the year 2003 described the motive of Information technology and the internet as follows: Presenting the data obtain ability and clarity in an easy manner; quantifying a delegation for data; Authorizing final actions which are based on the information provided by supply chains; allowing partners association [20]. Given below are the tables regarding the literature reviews done for the Indian automobile sector's three segments in detail briefing the

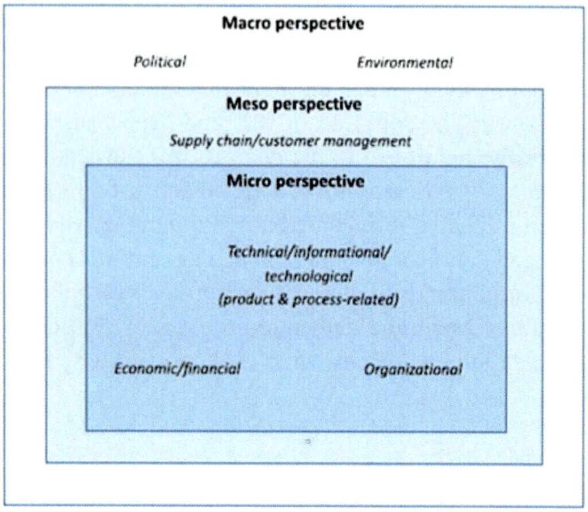

Fig. 1 Perspective of enablers

challenges faced by them while implementing AI in their supply chains at each level. A reference framework is drafted telling about a broader perspective of the enablers that support the implementation of AI in the Indian automobile supply chain system. The most basic or fundamental factors are categorized under the Micro perspective. It covers the most detailed factors like technical factors, economical factors, organizational factors, and other circumstances related to the product and processes. The next intermediate phase is termed the Meso phase where the broader factors like the supply chains and the end users or the customers come into the limelight. The last territorial perspective is termed as Macro perspective considers factors like political and environmental workspaces (Fig. 1).

2.1 Literature Related to Enablers of Commercial Vehicles

The motor vehicles which are used to carry goods, passengers from one location to another in order to earn profits for any corporation or any private personnel are termed as commercial vehicles. These include buses, trucks, trailers, coaches, etc. [21]. Commercial vehicles are some of the most heavy-duty vehicles on the road with complex gearing. Some of the major commercial vehicle manufacturers in India are Tata Motors, Ashok Leyland, Mahindra and Mahindra, Eicher Motors, Force Motors, etc. which are best examples of Indian Automobile SCM [22]. Table 1 gives the description of the enablers or the drivers and their effects that are faced by the various manufacturing stakeholders and end users while implementing AI in their supply chain systems. There are various green signals that can enhance the usage of AI technology in the supply chain systems which include top management support, financial dependence, skilled labor availability, etc. AI is an ongoing supporter of the continuous supply chain systems that consider various features as its success factors, i.e., skilled labor which is willing to update and upgrade with new innovations. Proper availability of infrastructures with well-maintained internet services are some of the main enablers. Table 1 describe the various enablers corresponding to the commercial vehicles in detail.

2.2 Literature Related to Enablers of Passenger Vehicles

Passenger cars are motor vehicles used for transporting passengers from one place to another, with a minimum of four wheels and consisting of a maximum of eight seats along with driver's space. For the carriage of goods vehicles with at least four wheels are categorized as light commercial vehicles [23]. The passenger vehicles include electric powered cars and vehicles, mini vans, sports utility vehicles, cars, vans, light trucks. Some of the major passenger vehicles manufacturers in India are Ford Cars India, Honda Cars India, Hyundai Motors, Jeep India Ltd, KIA Motors India, Audi India Ltd, Maruti Suzuki India, Skoda, Volvo Ltd., etc. [24]. After an intensive literature review in the Indian automobile sector of passenger vehicles division, the enablers faced by the different stakeholders were tabulated such as the enhanced sales and organizational motivation in the raw material supplier section. AI is a technology that needs a proper automated and well-structured setup for the best implementations. A motivational support by the top-level management members may enhance the implementation of AI in the SCM. The availability of properly trained and skilled workers in the component manufacturers becomes a pace maker while bringing AI into action. There are several instances where it is observed that several government policies in certain regions promote the technology along with the willingness of the workforce. In some other cases, it is studied that the customers or end users are getting adequate feedback from the authorities although using the technology. Table 2 details the enablers corresponding to different stakeholders of the passenger vehicles supply chains stakeholders while implementing AI in their systems.

2.3 Literature Related to Enablers of Two/Three Wheelers Vehicles

Any vehicle that runs on two wheels is called two-wheeler vehicles. Both the wheels may be placed in collaboration with each other and one behind the other, or arranged abreast on the common shaft whereas a vehicle equipped with three wheels, is termed as three-wheeler vehicle like auto rickshaws or a motorcycle with a sidecar. Some of the Indian manufacturing units of two-wheelers and three-wheelers are Hero Motocorp Ltd., Royal Enfield Pvt. Ltd., Java, Bajaj Auto Ltd., TVS Motor Company Ltd. [25]. The two-wheelers segment of the Indian automotive industry carries the

Table 1 Enablers corresponding to commercial vehicles

Stakeholder	Enablers	Effects	Citation
Raw material suppliers RMS	Top management support	Managerial support boosting the application of AI in the supply chains promotes the usage and indulgence of the technology	Kilpatrick and Factor (2000); LaLonde (2000); Monczka and Morgan (2017)
	Green signal from the government	Several government policies that come with new propaganda to make the economy digitalized are enhancers of AI	Chan and Kumar (2007); Balon et al. (2012)
	Good skilled workforce	Availability of skilled manpower that can handle the technology well in an automated manner	Ravi and Shankar (2005)
	Market demand	With the increased forecast when the workload increases on human labor the application of AI not only reduces the efforts but also increases the efficiency	Cleophas and Cleophas (2010); Kahn (2017)
	Acceptance by component manufacturer	With increasing demands when the component manufacturer and sub-assembler can easily place the upcoming consignments with due respect to the prevailing stock, AI is approved by the organization	Van Wijk, Jansen, and Lyles (2008); Tripathi, Rangarajan, and Talukder (2018)
Component manufacturers and sub assemblers	Willingness to update	The workers' urge to work in an atomized workstation is a booster of the technology. Where maximum work is compiled by atomization	Hosseini (2007); Sarkis (2009); Lettice, Wyat t, and Evans (2010)

(continued)

Table 1 (continued)

Stakeholder	Enablers	Effects	Citation
	Economic support	The financial and economic support to enhance the implementation of AI-driven system is mandatory as the initial cost of setup is bit higher	Luthra, Manju, Kumar, and Haleem (2010); Ho Johnny et al.(2009)
	Technological enhancement	With the advancement in the field of digitalization and industry 4.0, there are several manual operations that have changed to automated levels, with higher efficiency and accuracy	Sun and Medaglia(2019); AlKhidir&Zailani, (2009)
	Reverse logistics	Some of the products that have to be returned to the RMS due to the inapplicability of the material can be easily done once AI-driven systems are figured out	Sarkis (2012), José (2017); Batista et al. (2018)
Distribution channels	Financial factors	A smart monetary amount aided for the incorporation of AI by the top managerial authorities plays a vital role in technological enhancement	Jeyaraj, and M. C. Lacity (2006)
	Enhancing brand name	The company's brand name and market share value are hiked by the advancement in technology thereby giving better services at all levels	Yu Lin (2007); Yu Lin et al. (2008); Hsu et al. (2008); Chien et al. (2007)
	Good IT infrastructure	Provision of an appreciable IT infrastructure with all the facilities of networking helps the supply chain to be automated with AI	Sharma (2000); Stock (2001); Russell and Hoag (2004); Schwarz (2008);

(continued)

Table 1 (continued)

Stakeholder	Enablers	Effects	Citation
	Clean technology availability	An updated version of all the technological versions available will be beneficial for both the working environment and the workers	Zhu and Sarkis(2006); Li and Olorunniwo (2008); Giunipero et al. (2012)
End-user customers	Customer awareness	The interest shown by the end users along with consciousness to gain the best out of services provided without any hustle	Hosseini (2007); Yu Lin (2007); Liao, Hong, and Rao, (2010); Mudgal et al. (2010)
	Easy service availability	Trouble-free and undelayed services are provided to the customers just on the click of a button from their AI-driven systems linked with the companies	Wu et al. (2009); Ravi et al. (2005)
	Internet availability	The ease of availability of wifi and internet connection to the maximum of the population at nominal expenses tends to make the supply chain automated	P. Rao, and D. Holt (2005)

maximum number of purchase heads in the automobile market. Here the execution of AI in the supply chain system is vital for making the process automated and defect free [26]. The RMS being the primary stakeholder has great eco literacy among its workers along with top managerial support as the driving factor. The workforce in the component manufacturers and the sub-assemblers who are interested in changing their primitive ways of manual working are wholeheartedly welcoming to AI [27]. For setting up any new technology, there is a need for a compatible setup that enhances the working so are the funds required, so if a proper arrangement of the economic considerations is made along with well-equipped infrastructure that can act as a promoter for the technology [28]. Meanwhile customers these days are very aware of the usage of Artificial Intelligence that may even learn with the help of aiding applications. Table 3 details the enablers corresponding to the stakeholders and consumers of the two-wheeler/three-wheeler vehicles supply chains stakeholders while implementing AI in their systems.

Table 2 Enablers corresponding to passenger vehicles

Stakeholder	Enablers	Effects	Citation
Raw material suppliers RMS	Enhanced sales	With the increase in demand for raw materials in the market, there comes a need for automation	Yu Lin (2007); Yu Lin et al. (2008); Ravi et al. (2005)
	Political policies	With certain government policies that adorn digitalization and online billing the demand for the system to be artificially intelligence driven increases	Hosseini (2007);Mudgal et al. (2009); AlKhidir et al. (2009)
	Positive response	An affirmative retaliation from the above and the lower channels that aims at easing the working process with the aid of technology increases the utility	Van Wijk, Jansen, and Lyles (2008); Tripathi, Rangarajan, and Talukder (2018)
	Organizational motivation	The higher authorities' provocation when given in a gracious way the workers inclined to use the AI in their supply chains	Turker and Altuntas (2014); Horváth and Szabó (2019); Jabbour et al. (2018a)
Component manufacturers and sub-assemblers	Adopting innovation	The zeal of the co-workers is always ready to learn new modifications and new trending technology to upgrade their working abilities	Charan, Shankar, and Baisya, (2009); Novak and Stern (2008)
	Framework performances	The supply chains being a complex combination of various networks of workstations, its performance efficiency increases incredibly by the incorporation of AI	Sohal et al. (2001) and Kwan (1999)
	Sufficient data	With the availability of an ample amount of data set that is needed to start up the artificially intelligence supported systems, there are chances of high implications	Oliva and Watson (2009)
	Trust in AI in supply chain	The assurance and conviction of the AI in the ongoing supply chain process enhances the trust among the workers boosting their confidence about the accuracy and performances	Jin, Jeong & Kim (2017); Charan et al. 2009;
	Strategic planning	A well-defined step-wise plan for using AI, that can play a vital role in incorporating technology into the supply chains	Mudgal et al. (2009); Ravi et al. (2005); Zhu et al. (2004); Zhu et al.(2007); Zhu et al. (2008)

(continued)

Table 2 (continued)

Stakeholder	Enablers	Effects	Citation
Distribution channels	Economic considerations	The proper supply of funds and enumeration given to the distributors for using AI in their systems can increase proficiency	Mont et al. (2017), Preston (2012), Lin et al. (2013)
	Managerial support	The top-level management when supporting the system for using AI in their distribution channels the working capacity increases along with the rise in sales and profits	Mehrabi et al. (2012); Lee (2008); Wang et al. (2008); Wu et al. (2012);Luthra et al. (2013)
	Positive feedback	With the use of AI in the distribution channel an optimistic and efficacious response is received from the consumers and the sub-assemblers as their assignments are at ease	Reza Tizhoosh and Pantanowitz (2018)
	Educated workers	The skilled and educated workforce easily assimilates the new technological changes and accepts it for the benefits	Jeyaraj, and M. C. Lacity (2006)
End-user customers	Customer knowledge	The awareness created in the customers for the easy availability of the services and easy working boosts the application of AI	Mudgaletal. (2009); Ravi et al. (2005);Zhu et al. (2004); Zhu et al.(2007); Zhu et al. (2008)
	Time saver	AI not only increases the work speed but also saves the time thereby eradicating service delays	Mudgal et al. (2009); Mudgal et al. (2010); Zhu (2007)

3 Findings

After reviewing an ample number of research papers related to the field of implementing AI in the Indian automobile supply chains there are several outcomings that came from the past few decades of research. For any technology to be successful there are certain risk-taking factors that need to be approved and act as promoters, organizers, or facilitators. These not only help in the application of the technology but also drive the technology in a smooth manner keeping them as the driving force [29]. The research work done here was in a qualitative manner and by the adoption of certain qualitative assessment tools, some important findings were recognized. By reviewing more than 60–70 research works there were certain factors that kept on reiterating themselves as the driving enablers for AI. A comparison plot for the enablers with the amplitude of their occurring frequency was obtained. The area occupied by

Table 3 Enablers corresponding to two-/three-wheeler vehicles

Stakeholder	Enablers	Effects	Citation
Raw material suppliers RMS	High priority by management	Top preference and predominance by the managers that emphasize the system to be well operated by automation keep it as a key factor to be the enabler	Reza Tizhoosh and Pantanowitz (2018)
	Trust in technology	A deep faith in the field of AI that removes all the hurdles and myths about the technology keeping the perspective positive and motivating others for its operations	Turker&Altuntas (2014); Horváth and Szabó (2019); Jabbour et al. (2018a)
	Eco literacy among workers	Keeping in mind the ecological factors that harm the earth's ecology, i.e., the use of more paper and other resources, some workforces aim at making the system digitalized which not only saves the ecosystem but also improves the working	P. Rao, and D. Holt (2005)
	Government support	Some legal policies tend to make the systems digitalized and the government provides sustenance for the advancement of the technological enablers	Yu Lin (2007); Yu Lin et al. (2008); Ravi et al. (2005)
	Skilled workforce	Properly educated and skilled labor tends to learn the new progress and evolution of the updated technology	Khanna et al. (2013)
Component manufacturers and sub-assemblers	Increased profits	With the application of artificial intelligence in the manufacturing and assembling zone there is a hike in the financial gains and the bounces that are gained	Cleophas and Cleophas (2010) and Kahn (2017)
	Proper training provision	A well-planned training session on how to use the systems that are incorporated with AI for the subordinates gives the confidence in the work area	Lee and Whang (2000); Ayers, J.B. (2001)

(continued)

Table 3 (continued)

Stakeholder	Enablers	Effects	Citation
	Logistics	From the point of origin to the point of consumption the raw material in its semi-finished form transfers from different locations, storage, inventory, and processes. All these can be well handled by the AI applications	Sarkis (2012), José (2017); Batista et al. (2018)
	Sufficient IT infrastructure	Fully furnished and well-equipped state-of-the-art infrastructure workstations for AI-driven systems with round-the-clock internet availability	Katiyar and Barua (2013); Chandramowli et al. (2011)
Distribution channels	Increased market value	The sales and the profitability of the products increased by the AI-driven supply chains	Khanna et al. (2013)
	Awareness	The knowledge about the subject is so vastly increasing in the industry that much corporations are insisting to apply	Reza Tizhoosh and Pantanowitz (2018)
	Innovation	The transforming and upheaving the pre-existing methods of working in supply chains makes the need for implicating AI and automation	P. Rao, and D. Holt (2005)
	Higher efficiency	The supply chains that have AI applications in them have greater proficiency and competence as compared to manually operated supply chains	Turker and Altuntas (2014); Horváth and Szabó (2019); Jabbour et al. (2018a)
	Encouragement from the customers	When the customers demand the services that are automatically handled by artificial intelligence then it becomes the need of the hour	Cleophas and Cleophas (2010); Kahn (2017)
End-user customers	Customer interest	The engrossment that the customers possess for the use of automation in their after-sales services enhances the use of AI	Oliva and Watson (2009)

(continued)

Table 3 (continued)

Stakeholder	Enablers	Effects	Citation
	Easy and appreciable feedback	The trouble-free and ongoing critiques are obtained by the consumers by the AI driven systems by Chabot	Jeyaraj, and M. C. Lacity (2006)
	On door services	Just on single click of the button, the automated systems provide the on door services to the customers, reducing their movements to the agency thus satisfying the customer needs	Rogers et al. (1998); Wu et al. (2009); Ravi et al. (2004)

the individual factor depicts its degree of success in enabling AI in the Indian supply chains [30, 31]. As it is clear from the figure the maximum area of the box is covered by factors like top management support, political policies & IT infrastructure. These are followed by the lesser are that are covered by the enablers like skilled workforce, willingness to adopt new innovation and economic factors. Some vaguely affecting factors cover the smallest proportion like market demand, suppliers involvement, etc. Figure 2 describes the area-wise comparison chart of the success factors. Figure 3 depicts the top 10 enablers that were actually the main driving and success factors. The percentage-wise distributions of the nodes are demonstrated. Top-level managerial support and customer awareness are the greatest percentage holder factors which are followed by a skilled workforce and their willingness to change. The IT infrastructure and social considerations take the rear end in the percentage distribution. Figure 4 gives a graphical view of the top 5 enablers in percentage covered where the greatest is support from top-level management, followed by customer awareness, economic factors, IT infrastructure and hence the political forces. These enablers describes the qualitative aspect of the study.

4 Conclusions

The NVivo qualitative tool assessment embarks with the reiterating factors that were acting as the enablers of AI in Indian automobile supply chains. There are several factors that were identified as the main success factors or the drivers that caused the better incorporation of AI technology in the Indian Automobile supply chains. Certain enablers like the availability of proper IT infrastructure, economic considerations, customer awareness and a green flag from the top level management to boost the system to be automated by the AI technology creates a zone that deals with a good quality supply chain driven by AI. The error occurrence chances are reduced considerably. A hike in profits and working efficiency is witnessed along with increased sales and after-sale services. A higher level of customer satisfaction is obtained by the AI-driven systems rather than the manually operated systems. Even the graphical

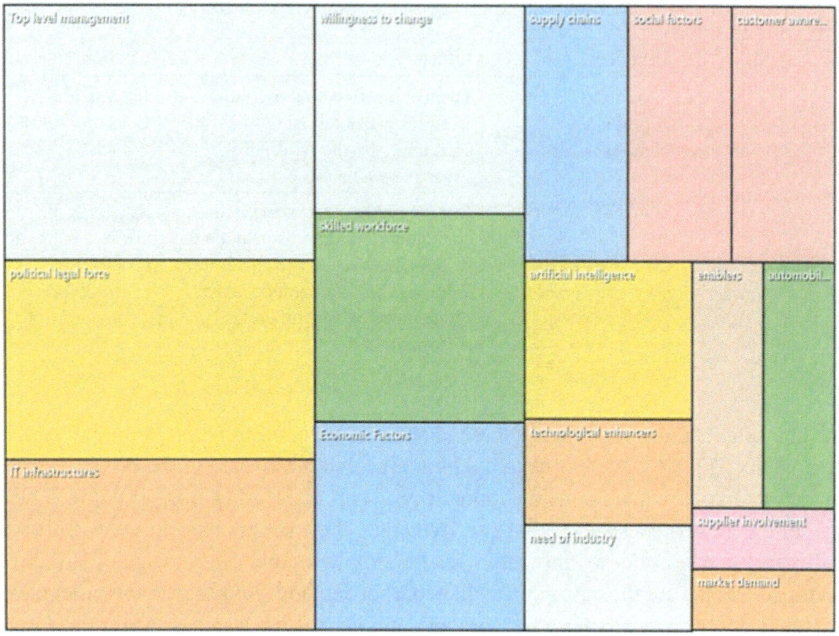

Fig. 2 Area comparison chart of the enablers

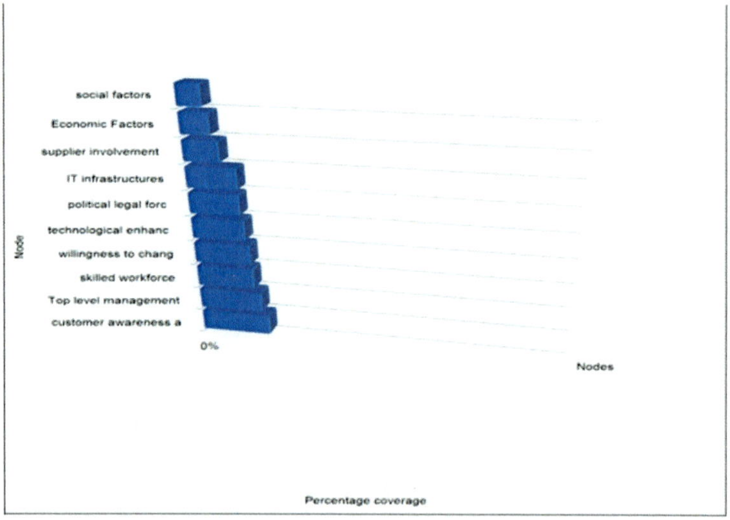

Fig. 3 Node comparison percentage wise

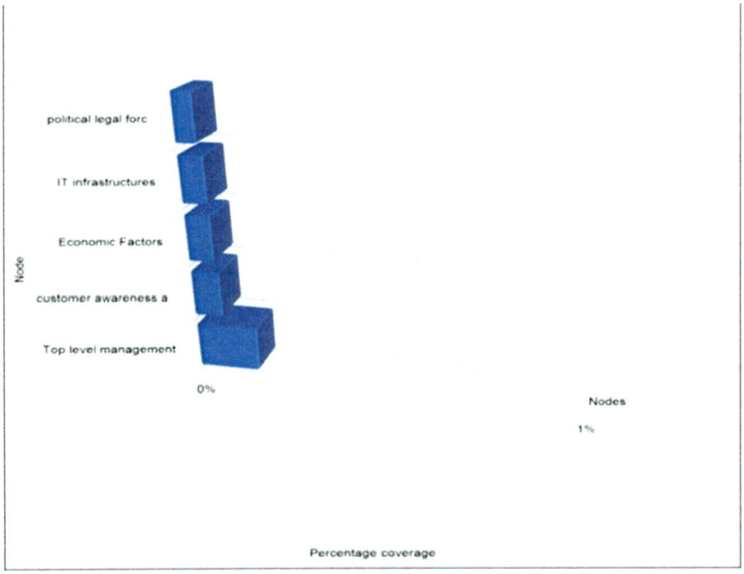

Fig. 4 Percentage coverage of top five nodes or enablers

representations of the success factors obtained by the qualitative assessment tool show the degree of extent of function ability of the AI-driven automobile supply chains.

References

1. Ansari ZN, Kant R (2017) A state-of-art literature review reflecting 15 years of focus on sustainable supply chain management. J Clean Prod 142:2524–2543. https://doi.org/10.1016/j.jclepro.2016.11.023
2. Dumitrascu O, Dumitrascu M, Dobrotă D (2020) Performance evaluation for a sustainable supply chain management system in the automotive industry using artificial intelligence. Processes 8(11):1384. https://doi.org/10.3390/pr8111384
3. Luthra S, Mangla SK (2018) Evaluating challenges to industry 4.0 initiatives for supply chain sustainability in emerging economies. Process Saf Environ Prot 117:168–179. https://doi.org/10.1016/j.psep.2018.04.018. Gandhi, MA (2017) A review of the Indian automobile manufacturing sector. IOSR J Bus Manag 19(03):09–15. https://doi.org/10.9790/487X-1903020915
4. Jharkharia and Shankar (2005) IT-enablement of supply chains understanding the industry. In: Proceeding of international conference of engineers and computers scientist, vol 3, March 17–19
5. Kamble SS, Gunasekaran A, Sharma R (2018) Analysis of the driving and dependence power of barriers to adopt industry 4.0 in Indian manufacturing industry. Comput Ind 101:107–119. https://doi.org/10.1016/j.compind.2018.06.004
6. Borade AB, Bansod SV (2007) Domain of supply chain management—a state of art. J Technol Manag Innov 2(4):109–121. ISSN: 0718-2724. www.jotmi.org

7. Jharkharia S, Shankar R (2005) IT enablement of supply chains: understanding the barriers. J Enterp Inf Manag 18(1):11–27
8. Kumar Sharma S, Bhat A (2014) Supply chain risk management dimensions in Indian automobile industry: a cluster analysis approach. Benchmarking: Int J 21(6):1023–1040. https://doi.org/10.1108/BIJ-02-2013-0023
9. Min H (2010) Artificial intelligence in supply chain management: theory and applications. Int J Log Res Appl 13(1):13–39. https://doi.org/10.1080/13675560902736537
10. Mostafa M (n.d.) Artificial intelligence applications in supply chain management and analysis in Turkey. Int J Manag Decis Mak 9(2):163–182
11. Olugu EU, Wong KY, Shaharoun AM (2010) A comprehensive approach in assessing the performance of an automobile closed loop supply chain. Sustainability 2:871–879
12. Dumitrascu O, Dumitrascu M, Dobrotă D (2020) Performance evaluation for a sustainable supply chain management system in the automotive industry using artificial intelligence. Processes 8(11):1384. https://doi.org/10.3390/pr8111384
13. Peres RS, Jia X, Lee J, Sun K, Colombo AW, Barata J (2020) Industrial artificial intelligence in industry 4.0—systematic review challenges and outlook. IEEE Access 8:220121–220139. https://doi.org/10.1109/ACCESS.2020.3042874
14. Safar L, Sopko J, Dancakova D, Woschank M (2020) Industry 4.0—awareness in South India. Sustainability 12(8):3207. https://doi.org/10.3390/su12083207
15. Kamble S, Gunasekaran A, Dhone NC (2020) Industry 4.0 and lean manufacturing practices for sustainable organisational performance in Indian manufacturing companies. Int J Prod Res 58(5):1319–1337. https://doi.org/10.1080/00207543.2019.1630772
16. Frank AG, Dalenogare LS, Ayala NF (2019) Industry 4.0 technologies: implementation patterns in manufacturing companies. Int J Prod Econ 210:15–26. https://doi.org/10.1016/j.ijpe.2019.01.004
17. Rehman, M. A. A., Aneyrao, T. A., &Shrivastava, R. L. (2015). Identification of critical success factors in Indian automobile industry: A GSCM approach. International Journal of Process Management and Benchmarking, 5(2), 229. https://doi.org/10.1504/IJPMB.2015.068670
18. Srinivas, S. S., &Sreedharan, V. R. (n.d.). Failure analysis of automobile spares in a manufacturing supply chain distribution centre using Six Sigma DMAIC framework. 14.
19. Susairaj Dr AX, Salaijayamani A, Premkumar A (2021) Impact of Covid-19 on Automobile Industries in India. Emperor J Econ Soc Sci Res 3(05):40–50. https://doi.org/10.35338/EJESSR.2021.3505
20. Toorajipour R, Sohrabpour V, Nazarpour A, Oghazi P, Fischl M (2021) Artificial intelligence in supply chain management: a systematic literature review. J Bus Res 122:502–517. https://doi.org/10.1016/j.jbusres.2020.09.009
21. Tripathi, Talukder (2018) 2020 Supply chain performance and profitability in India. Process Saf Environ Prot 117:168–179. https://doi.org/10.1016/j.psep.2018.04.018
22. Tripathi S, Talukder B (2020) Supply chain performance and profitability in Indian automobile industry: evidence of segmental difference. Glob Bus Rev:097215091989830.https://doi.org/10.1177/0972150919898302
23. Albers A, Gladysz B, Pinner T, Butenko V, Stürmlinger T (2016) Procedure for defining the system of objectives in the initial phase of an industry 4.0 project focusing on intelligent quality control systems. Procedia CIRP 52:262–267. https://doi.org/10.1016/j.procir.2016.07.067
24. Mittal S, Khan MA, Romero D, Wuest T (2018) A critical review of smart manufacturing & industry 4.0 maturity models: implications for small and medium-sized enterprises (SMEs). J Manuf Syst 49:194–214. https://doi.org/10.1016/j.jmsy.2018.10.005
25. Motyl B, Baronio G, Uberti S, Speranza D, Filippi S (2017) How Will Change the Future Engineers' Skills in the Industry 4.0 Framework? A Questionnaire Survey. Procedia Manuf 11:1501–1509. https://doi.org/10.1016/j.promfg.2017.07.28
26. Büyüközkan G, Çifçi G (2012) A novel hybrid MCDM approach based on fuzzy DEMATEL, fuzzy ANP and fuzzy TOPSIS to evaluate green suppliers. Expert Syst Appl 39(3):3000–3011. https://doi.org/10.1016/j.eswa.2011.08.162

27. Farooque M, Jain V, Zhang A, Li Z (2020) Fuzzy DEMATEL analysis of barriers to Blockchain-based life cycle assessment in China. Comput Ind Eng 147:106684. https://doi.org/10.1016/j.cie.2020.106684
28. Govindan K, Kannan D, Mathiyazhagan K, Jabbour ABL de S, Jabbour CJC (2013) Analysing green supply chain management practices in Brazil's electrical/electronics industry using interpretive structural modelling. Int J Environ Stud 70(4):477–493.https://doi.org/10.1080/00207233.2013.798494
29. Kumar S, Raut RD, Nayal K, Kraus S, Yadav VS, Narkhede BE (2021) To identify industry 4.0 and circular economy adoption barriers in the agriculture supply chain by using ISM-ANP. J Clean Prod 293:126023. https://doi.org/10.1016/j.jclepro.2021.126023
30. Singh BJ, Bakshi Y (2014) Optimizing backup power systems through Six Sigma: an Indian case study of diesel genset. Int J Lean Six Sigma 5(2):168–192. https://doi.org/10.1108/IJLSS-09-2012-0008
31. Singh BJ, Joshi R (2015) Benchmarking academics through sustainable assessment criteria: an Indian case study. Benchmarking: Int J 22(3):505–536.https://doi.org/10.1108/BIJ-01-2013-0006

Evaluation of Ergonomic Assessment Tools Using Fuzzy AHP

Swattvik som and Lakhwinder Pal singh

Abstract Work-related Musculoskeletal Disorders (WMSDs) are on the rise in almost every industrialized country. For countering the effects of WMSDs posture analysis tools are used by ergonomic practitioners. The collection of posture analysis tools is huge, with the likes of RULA, REBA, QEC, OCRA, JSI, and others. Here comes the use of the Fuzzy Analytic Hierarchical Process (F-AHP). Recently, F-AHP is used extensively for the selection of optimal option out of a set of choices. F-AHP is primarily based on the Analytic Hierarchy Process (AHP), it is a Multi-Criteria Decision-Making (MCDM) model, which uses the Fuzzy theory. In this research, F-AHP is used to evaluate five posture analysis tools, viz. RULA, REBA, QEC, OCRA, and JSI. A hierarchical structure was constructed by considering Work posture, Frequency/Repetition, and Load/Weight handled as factors (level 1) and RULA, REBA, QEC, OCRA, and JSI as the choices (level 2). The available epidemiologic evidence and views of experts were prime drivers in conducting this research. MS Excel was used to conduct the requisite calculations of AHP. CR was found to be less than 0.10 in each case, suggesting that there is no anomaly in consistency. This research led to the conclusion that QEC was overall best among the tools selected, followed by JSI, REBA, RULA, and OCRA respectively.

Keywords Fuzzy AHP · RULA · REBA · OCRA · JSI · QEC

1 Introduction

WMSDs affect various parts of the body like the cartilage, tendons, nerves, ligaments, muscles, and joints and include symptoms including pain, inflammation, tingling, and numbness. Musculoskeletal Disorders (MSDs) that are brought on or made worse by work are referred to as WMSDs [1]. Such a group of MSDs has a huge impact on

S. som (✉) · L. P. singh
Department of Industrial and Production Engineering, Dr B R Ambedkar Natrional Institute of Technology, Jalandhar 144027, India
e-mail: sswattvik@gmail.com

A. Sachdeva et al. (eds.), *Recent Advances in Operations Management and Optimization*, Lecture Notes in Mechanical Engineering,
https://doi.org/10.1007/978-981-99-7445-0_19

the employees and might take months or years to develop [2]. These are among the main causes of employees' absences from work, disabilities, medical expenses, and compensation claims [2]. Low back pain, cervical spondylosis, and osteoarthritis are a few of the typical MSDs. These illnesses cost roughly 23% of the entire cost of the sickness [3]. Even though most companies now choose mechanization or automation, workers in several sectors still need to do physical material handling duties to support their families. The most significant risk factors associated with these tasks are (a), uncomfortable postures (bending and twisting) (b) forceful exertions (heavy lifting as well as carrying of loads), (c) repetitive movements (regular lifting and carrying of loads), and (d) long-term static postures [4]. Any of the risk factors mentioned above may cause pain and weariness with repeated or prolonged exposure [5], but if the condition worsens, WMSDs may develop as a consequence. The length of time a worker is exposed, how often they are exposed to risk variables, and how intensely they are exposed all affect the amount of risk [1]. Employees are more susceptible over time to neck, back, shoulder, hand, wrist, and other types of ailments [6].

The greatest morbidity costs of all illnesses are likewise associated with MSDs. Most developed nations have recorded comparable numbers [7]. As per the World Health Organization (WHO), the major cause of work-related impairment in developed countries is MSDs [8]. India is now home to approximately 12% of the world's elderly population. According to demographic trends, it is predicted that between 2000 and 2050, the number of Indians in their 60s and older will expand by 326%, while the population of Indians 80 years and older will increase by 700%, making it the fastest-growing category [9]. With the growing elderly population, the MSDs will increase which contributes to the country's expenses [10]. In India, around 40% of the total population is engaged in industrial work. In certain industries of the workforce, ergonomic concerns are thought to be responsible for more than 50% of all MSDs [11]. The burden would be far greater and rapidly rise to unprecedented levels if MSD estimates were established for these two target groups. The impact of MSDs has received less attention than that of other non-communicable illnesses (NCDs). However, rapid demographic changes and epidemiological changes in sickness have led to a rise in the proportion of disease burden related to NCDs.

The risk factors related to WMSDs are repetition, excessive force application [12], vibration, and awkward postures [13]. Based on strong evidence, research shows a direct correlation between these risk factors and the occurrence of MSDs [14]. Moreover, psychosocial factors also play a role in the development of these injuries [15]. The neck, lower back, wrists, shoulders, elbows, forearms, and hands are the most commonly impacted areas [16]. In recent years, there are many published techniques taking interest in this context [17] including such as Rapid Upper Limb Assessment (RULA) [18], Rapid Entire Body Assessment (REBA) [19], Quick Exposure Check (QEC) [17], Job Strain Index (JSI) [20], Occupational Repetitive Action (OCRA) [21], etc.

The strategic aim of the study is to assess and find out the best fit ergonomic assessment tool among the five most extensively used ergonomic assessment tools namely RULA, REBA, QEC, OCRA, and JSI by using AHP.

2 Methodology

Initially, AHP was developed by Saaty for MCDM. Later on, fuzzy theory was blended with AHP and it resulted in the birth of F-AHP. AHP employs pair-wise comparisons of diverse choices in relation to a variety of criteria. The objective is at level 0 in the fundamental AHP model, whereas the criterion is placed at level 1 and the sub-criterion is placed at level 2, respectively. Finally, the options or choices are placed at level 3 [22]. While traditional AHP does not account for inconsistency in human judgments, it has been improved by amalgamating fuzzy logic. The linguistic variables, represented as triangular numbers [22] in F-AHP, are used to perform pairwise assessments of both, criteria and alternatives. Historically, F-AHP's application was developed and showcased by Van Laarhoven and Pedrycz [23]. For pairwise comparisons, the authors have defined Triangle Membership Functions (TMF). Following this, Buckley [24] contributed by recognizing the fuzziness of comparison ratios with TMF. A fresh approach involving the use of TFN in pair-wise comparisons was also presented by Chang [25]. Even though F-AHP contains more techniques, in the context of this study, Buckle's methods [24] are used to calculate the comparative relevance weights for the criteria and the choices. The following steps are used in the procedure (Table 1):

Step 1: Define the problem of the current study.

Step 2: Build the hierarchy in the top-down approach, where the top slot is taken as the goal of the study which is followed by factors and choices, respectively.

Step 3: Linguistic terms are used by experts to compare alternatives and factors, and based on the inputs of experts, triangular fuzzy sets are created and later on used to construct the set of pairwise comparison matrices (size $n \times n$). The number of decisions needed to create the set of matrices in step 3 is $n(n - 1)/2$.

These linguistic concepts have TFNs that correlate to them. For instance, if a professional claims that "Criterion 1 (C1) is Fairly Important than Criterion 2 (C2),"

Table 1 The triangular fuzzy numbers associated with linguistic words

AHP scale	Definition	TFN
1	Equal preference	(1, 1, 1)
2	Intermediate between equal and moderate	(1, 2, 3)
3	Moderate preference	(2, 3, 4)
4	Intermediate between moderate and strong	(3, 4, 5)
5	Strong preference	(4, 5, 6)
6	Intermediate between strong and very strong	(5, 6, 7)
7	Very strong preference	(6, 7, 8)
8	Intermediate between very strong and extremely preferred	(7, 8, 9)
9	Extremely preferred	(9, 9, 9)

then on a triangular fuzzy scale (TFS), it is represented as (4, 5, 6). In contrast, the comparison of C2 to C1 in the pairwise contribution matrix of the criteria will use the TFS of (1/6, 1/5, 1/4). In Eq. 1, the pairwise contribution matrix is displayed, and $\tilde{a}_{ij}{}^{k}$ denotes, via TFN, the kth decision maker's preference for the *i*th criterion over the *j*th criterion. Here, the "tilde" represents the triangular number demonstration.

$$\tilde{A}^{k} = \begin{bmatrix} \tilde{a}_{11}^{k} & \cdots & \tilde{a}_{1n}^{k} \\ \vdots & \ddots & \vdots \\ \tilde{a}_{n1}^{k} & \cdots & \tilde{a}_{nn}^{k} \end{bmatrix} \tag{1}$$

Step 4: In the case of multiple decision makers the respective preferences are averaged and hence updated by using Eq. 2.

$$\tilde{a}_{ij} = \frac{\sum_{k=1}^{k} \tilde{a}_{ij}^{k}}{K} \tag{2}$$

Step 5: Geometric mean of the comparison values, which are fuzzy in nature, are calculated using Eq. 3. Here, $\tilde{u}_i$ still represents triangular values [24].

$$\tilde{u}_i = \left(\prod_{j=1}^{n} \tilde{a}_{ij} \right)^{1/n} \quad i = 1 \text{ to } n \tag{3}$$

Step 6: Equation 4 is used to calculate the fuzzy weights of each criterion, by following the below 3 sub-steps.

Step 6.1: For each $\tilde{u}_i$, the vector sum is calculated.

Step 6.2: Inverse of summation vector is found out. Replace the TFN, so as to arrange it in an increasing order.

Step 6.3: Each $\tilde{u}_i$ is multiplied with the above-mentioned reverse vector to find the fuzzy weight of criteria i ($\tilde{w}_i$)

$$\begin{aligned} \tilde{w}_i &= \tilde{u}_i * (\tilde{u}_1 * \tilde{u}_2 * \cdots * \tilde{u}_n)^{-1} \\ &= (ew_i, fw_i, gw_i) \end{aligned} \tag{4}$$

Step 7: The fuzzy nature of $\tilde{w}_i$ is de-fuzzified by the centre of area method proposed by Chou and Chang [26], by using Eq. 5.

$$Mi = \frac{ew_i + fw_i + gw_i}{3} \tag{5}$$

Step 8: Afterwards, Mi is normalized by following Eq. 6.

$$\mathrm{Ni} = \frac{\mathrm{Mi}}{\sum_{i=1}^{n} \mathrm{Mi}} \tag{6}$$

Step 9: After completion of all the pairwise comparisons, the consistency is determined by using the Eigenvalue, λ_{max} to calculate the consistency index in the following way:

$$\mathrm{CI} = \frac{\lambda_{max} - n}{(n - 1)} \tag{7}$$

where n is the matrix size. The consistency of the judgment can be checked by calculating the consistency ratio (CR) by taking an appropriate value of CI as mentioned in Table 2. The acceptable limit of CR is ≤ 0.10. If it is more, the judgment matrix is inconsistent. The judgments should be reviewed and improved accordingly to obtain a consistent matrix.

To determine the normalized weights of both criteria and options, these nine procedures are taken. The scores for each alternative are then computed by dividing each alternative weight by the required criteria. These findings indicate to the decision maker the alternative with the greatest score. In this paper, 5 ergonomic assessment tools were selected under the banner of choices (level 2), namely RULA, REBA, QEC, OCRA, and JSI as these tools were found to be the most commonly used by ergonomists for assessment [27]. The factors chosen here were work posture, frequency/repetition, and load/weight handled since these 3 factors were found to be the primary cause of Work-Related Musculoskeletal Disorder (WMSD) [28]. The hierarchy was developed as shown in Fig. 1. A questionnaire was created and data was gained from 6 experts who are professors and Ph.D. scholars from reputed Indian institutes.

Table 2 Priority matrix for five ergonomic assessment tools

	Posture (0.42459)	Weight (0.42459)	Frequency (0.15082)	Overall priority
OCRA	0.06485	0.15986	0.12527	0.11430
JSI	0.03329	0.36165	0.50442	0.24376
RULA	0.26661	0.07755	0.05417	0.15430
REBA	0.50337	0.03930	0.05417	0.23858
QEC	0.13188	0.36165	0.26197	0.24906

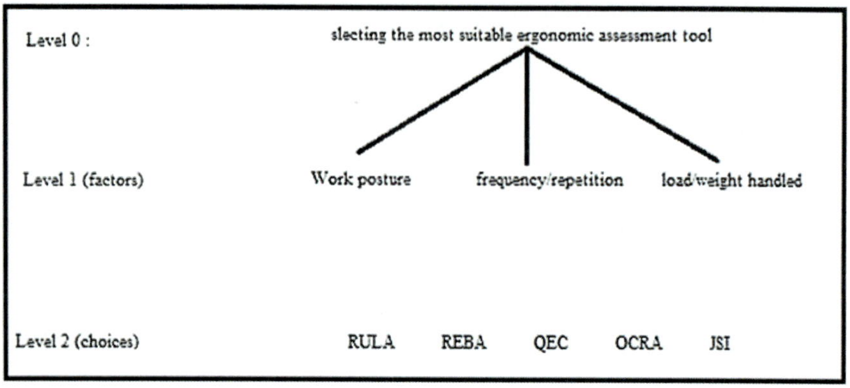

Fig. 1 Hierarchy for selection of best fit ergonomic assessment tool

3 Result and Conclusion

After conducting the AHP analysis of five ergonomic assessment tools the overall priority was found which is summarized in Table 2. It has been found that QEC obtained an overall priority score of 0.24906, followed by JSI (0.24376), REBA (0.23858), RULA (0.15430), and OCRA (0.11430). CR was found to be less than 0.10 in all the cases, thus suggesting that the judgments were consistent. From this study, it is concluded that QEC is the best option when dealing with a generic ergonomic assessment. QEC as a tool is also very user-friendly as it uses verbal assessments extensively during assessment which is a rather easier approach than the cumbersome process of studying angles and then scoring based on those angles.

In the future, this approach can be used with a different set of ergonomic assessment tools as well as a different set of factors depending upon workplace requirements and other technical reservations to conduct the ergonomic assessment.

References

1. Lei L, Dempsey PG, Xu J, Ge L, Liang Y (2005) Risk factors for the prevalence of musculoskeletal disorders among Chinese foundry workers. Int J Ind Ergon 35:197–204
2. Kuorinka I, Forcier L, Hagberg M, Silverstein B, Wells R, Smith MJ, Hendrick HW, Carayon P, Pérusse M (1995) Work related musculoskeletal disorders (WMSDs): a reference book for prevention
3. Jacobson L, Lindgren B, Sjukdomarna VK (1996) What are the costs of illness? Stock. Socialstyrelsen (National Board Health and Welfare)
4. Rossi D, Bertoloni E, Fenaroli M, Marciano F, Alberti M (2013) A multi-criteria ergonomic and performance methodology for evaluating alternatives in "manuable" material handling. Int J Ind Ergon 43:314–327

5. Van Der Beek AJ, Hoozemans MJM, Frings-Dresen MHW, Burdorf A (1999) Assessment of exposure to pushing and pulling in epidemiological field studies: an overview of methods, exposure measures, and measurement strategies. Int J Ind Ergon 24:417–429
6. Ergonomic guidelines for manual material handling (2020). https://doi.org/10.26616/NIOSHPUB2007131
7. Punnett L, Wegman DH (2004) Work-related musculoskeletal disorders: the epidemiologic evidence and the debate. J Electromyogr Kinesiol 14:13–23
8. Diet, nutrition and the prevention of chronic diseases. Switzerland (2003)
9. Jeyalakshmi S, Chakrabarti S, Gupta N (2011) Situation analysis of the elderly in India. Central Statistical Office (CSO), Ministry of Statistics and Programme Implementation, Government of India Doc
10. Leveille SG (2004) Musculoskeletal aging. Curr Opin Rheumatol 16:114–118
11. Back L (2001) Musculoskeletal disorders and the workplace
12. Silverstein BA, Fine LJ, Armstrong TJ (1987) Occupational factors and carpal tunnel syndrome. Am J Ind Med 11:343–358
13. Pope MH, Magnusson M, Wilder DG (1998) Low back pain and whole body vibration. Clin Orthop Relat Res 354:241–248
14. Grieco A, Molteni G, de Vito G, Sias N (1998) Epidemiology of musculoskeletal disorders due to biomechanical overload. Ergonomics 41:1253–1260
15. Punnett L, Bergqvist U (1997) Visual display unit work and upper extremity musculoskeletal disorders. Stock Natl Inst Work Life 997
16. Buckle PW, Devereux JJ (2002) The nature of work-related neck and upper limb musculoskeletal disorders. Appl Ergon 33:207–217
17. David G, Woods V, Li G, Buckle P (2008) The development of the quick exposure check (QEC) for assessing exposure to risk factors for work-related musculoskeletal disorders. Appl Ergon 39:57–69
18. McAtamney L, Corlett EN (1993) RULA: a survey method for the investigation of work-related upper limb disorders. Appl Ergon 24:91–99
19. Hignett S, McAtamney L (2000) Rapid entire body assessment (REBA). Appl Ergon 31:201–205
20. Jones T, Kumar S (2010) Comparison of ergonomic risk assessment output in four sawmill jobs. Int J Occup Saf Ergon 16:105–111
21. Colombini D (1998) An observational method for classifying exposure to repetitive movements of the upper limbs. Ergonomics 41:1261–1289
22. Kilincci O, Onal SA (2011) Fuzzy AHP approach for supplier selection in a washing machine company. Expert Syst Appl 38:9656–9664
23. Van Laarhoven PJM, Pedrycz W (1983) A fuzzy extension of Saaty's priority theory. Fuzzy Sets Syst 11:229–241
24. Buckley JJ (1985) Fuzzy hierarchical analysis. Fuzzy Sets Syst 17:233–247
25. Chang D-Y (1996) Applications of the extent analysis method on fuzzy AHP. Eur J Oper Res 95:649–655
26. Chou S-W, Chang Y-C (2008) The implementation factors that influence the ERP (enterprise resource planning) benefits. Decis Support Syst 46:149–157
27. Joshi M, Deshpande V (2019) A systematic review of comparative studies on ergonomic assessment techniques. Int J Ind Ergon 74:102865
28. The Definition and Causes of Musculoskeletal Disorders. https://ergo-plus.com/musculoskeletal-disorders-msd/

Electric Vehicles: Consumer Perceptions and Expectations

Atul Zope, Raju Kumar Swami, and Atul Patil

Abstract Electric vehicles (EVs) have emerged as the future of the automotive industry. Technological improvements have further reduced the cost of EVs making them more economically viable and an environmentally friendly alternative to IC engine-based vehicles. However, there are several parameters that influence consumers' decisions to purchase an EV. This article explores these parameters and the impact they have on consumers' perceptions and expectations of EVs. A survey was conducted for various economic, technological, social, and environmental aspects to understand the perceptions related to electric vehicle acceptance and its expectations. The paper discusses the results of a survey conducted to analyze the various factors that influence consumer perceptions and expectations of electric vehicles (EVs). The survey results indicate that the cost of purchasing an EV, insurance affordability, increasing price of petrol and diesel, mileage (energy consumption), EV incentives, and battery replacement/repair cost have a moderate to very strong impact on consumer decisions. The results also show that the cost of maintenance, time required for maintenance, and ease of maintenance have a moderate impact on consumer perceptions and expectations. The survey results further indicate that the charging time and availability of charging infrastructure have a moderate and low impact, respectively, on consumer decisions. The findings of this survey can be useful for policymakers, manufacturers, and other stakeholders to develop strategies to increase the adoption of EVs.

Keywords Electric vehicles · Battery · Maintenance · Economic aspects · Performance · Consumer expectations

A. Zope (✉) · R. K. Swami
Faculty of Engineering, Pacific Academy of Higher Education and Research University, Udaipur, India
e-mail: zope.atul@gmail.com

R. K. Swami
e-mail: raju.swami@pacific-it.ac.in

A. Patil
Department of Mechanical Engineering, D.Y. Patil Institute of Technology, Pimpri, Pune, Maharashtra, India

A. Sachdeva et al. (eds.), *Recent Advances in Operations Management and Optimization*, Lecture Notes in Mechanical Engineering,
https://doi.org/10.1007/978-981-99-7445-0_20

Nomenclature

EV	Electric Vehicle
FAME	Faster Adoption and Manufacturing of Hybrid and Electric Vehicles
GST	Goods and Service Tax
IC	Internal Combustion

1 Introduction

The internal combustion engine vehicles we have today pose a major threat to the environment and climate. Additionally, it is not a sustainable approach considering energy requirements. To address this, many countries, both developed and developing, are adopting electric vehicle technology. Governments are supporting this shift through various incentives such as tax breaks, lower loan interest rates, and free or subsidized public parking. In India, the Faster Adoption and Manufacturing of Hybrid and Electric Vehicles (FAME) scheme promotes electric mobility. Electric vehicles offer numerous benefits compared to internal combustion engines, such as zero emissions, low maintenance, high reliability, quiet and smooth rides, and better overall efficiency [1–3]. The cost of batteries has significantly decreased in the past five years, making EVs more affordable. However, there are some drawbacks, such as limited driving range, long charging time, and heavy battery packs that take up space and add weight to the vehicle. EVs provide a smoother and quieter driving experience due to reduced vibrations and the absence of engine noise [2].

2 Literature Survey

A literature survey is conducted by exhaustively considering all the parameters which may potentially affect the electric vehicle market.

2.1 Economic Aspects

Several papers have been reviewed, and it has been identified that the economic aspect is essential for the penetration of EV ideas in large markets of consumers. In order to make it more affordable to the consumers, various studies have been reviewed about the economic aspect and their critical findings have been mentioned in the result and discussion. Several economic aspects have been identified; namely purchase price, insurance affordability, increasing price of petrol and diesel, mileage (energy consumption), EV incentive—reduced GST, battery replacement/repair Cost [1, 4–16].

2.2 Maintenance

Several papers have been reviewed, and it has been identified that the maintenance aspect is essential for the penetration of EV ideas in large markets of consumer. In order to make it more affordable to the consumers, various studies have been reviewed about the economic aspect and their critical findings have been mentioned in the result and discussion. Several economic aspects have been identified; namely cost of maintenance, time of maintenance, ease of maintenance [17].

2.3 Battery and Charging Aspects

Several papers have been reviewed, and it has been identified that the economic aspect is essential for the penetration of EV idea in large markets of consumers. In order to make it more affordable to consumers, various studies have been reviewed about battery and charging aspects and their critical findings have been mentioned in the results and discussion. Several economic aspects have been identified; namely charging time, availability of charging infrastructure, battery life, battery warranty, battery capacity, battery overheating problem [14, 18–22].

2.4 EV Aspects

Several papers have been reviewed, and it has been identified that EV Aspects is essential for the penetration of EV idea in large market of consumer. In order to make it more affordable to the consumers, various studies have been reviewed about the economic aspect and their critical findings have been mentioned in the results and discussion. Several economic aspects have been identified; namely EV warranty, after-sales service and support, vehicle weight, driving range, aesthetics, software update, hardware update, environmental friendly vehicle, safety considerations [15, 22–25].

2.5 EV Performance Aspects

Several papers have been reviewed, and it has been identified that EV Performance Aspects is essential for the penetration of EV idea in large market of consumer.

In order to make it more affordable to the consumers, various studies have been reviewed about the economic aspect and their critical findings have been mentioned in the results and discussion. Several economic aspects have been identified; namely power, acceleration, torque, top speed [16, 26].

3 Methodology

Surveys can provide valuable information on complex and emerging issues and help decision-makers make informed decisions. The survey is a research method in which a sample of experts in a particular field is selected and asked to provide their opinions, insights, and predictions on a given topic. The methodology typically involves the following steps:

1. Define the research question and identify the target population of experts.
2. Select a sample of experts to participate in the survey. This can be done through various sampling methods, such as random sampling, purposive sampling, or snowball sampling.
3. Develop a questionnaire or interview guide to collect data from the experts.
4. Administer the survey, either through online platforms, phone, email, or in-person interviews.
5. Analyze the data collected and interpret the results.
6. Report the findings, highlighting the opinions, insights, and predictions of the experts on the research question.

The information regarding consumer perceptions and expectations of electric vehicles was collected through a literature survey and also through a survey of 196 participants involving engineers, technicians, consumers, and other professionals working in the electric vehicles domain. The survey was taken on a Likert scale of 1 to 5 where A type of response scale in which responders specify their level of agreement to a statement typically in five points: (1) very low impact; (2) low impact; (3) moderate impact; (4) strong impact; (5) very strong impact.

4 Results and Discussions

4.1 Economic Aspects

4.1.1 Purchase Price: The cost of purchasing an EV is one of the major factors that influence consumers' decisions, which is also evident from the survey where a moderate to very strong impact of EV purchase price is observed. Despite the numerous benefits of EVs, their higher price compared to traditional gasoline vehicles remains a significant barrier to entry for many consumers. However, with the increase in demand and the economies of scale, the cost of EVs is expected to decrease over time, making them more accessible to consumers. The cost of EVs has significantly decreased in recent years. The power train expenditure is estimated to reduce by up to 70% by 2030 [27]. One study found that the life-cycle environmental impact of Li-ion batteries accounts for about 15% of the overall impact of a BEV [28]. The sale of EVs is expected to continue to grow as more countries move towards banning internal combustion engine vehicles [4–10].

4.1.2 Insurance Affordability: From the survey, a moderate impact of insurance affordability is observed. The cost of ensuring an EV is another factor that affects consumer perceptions and expectations. As EVs are relatively new, the cost of insuring them can be higher compared to traditional gasoline vehicles. However, as the market for EVs grows and insurance companies become more familiar with these vehicles, it is expected that the cost of insurance will decrease. EV insurance is more expensive due to the high cost of batteries, especially for larger EVs. Repairing an EV can also be costly as they are more prone to damage [11].

4.1.3 Increasing Price of Petrol and Diesel: From the survey, a moderate to very strong impact of increasing price of petrol and diesel is observed. The increasing price of petrol and diesel is another factor that affects consumer perceptions and expectations of EVs. With the cost of gasoline and diesel rising, the cost of owning and operating an internal combustion engine vehicle becomes increasingly expensive. This makes EVs an attractive alternative, as they are cheaper to operate and maintain over the long term. With petrol and diesel prices at approximately Rs. 100 and 90 per litre, respectively, consumers are more likely to adopt EVs due to reduced maintenance and operating costs [12]. An EV costs approximately Rs. 1.3 per kilometre, compared to Rs. 10 for gasoline cars [13].

4.1.4 Mileage (Energy Consumption): From the survey, a moderate to very strong impact of mileage is observed. The energy consumption of EVs is a crucial factor that influences consumer perceptions and expectations. Consumers want to know how far they can travel on a single charge and the cost of recharging their EVs. With advances in battery technology, the driving range of EVs is increasing, and the cost of recharging is becoming more affordable. The impact of economic aspects on the electric vehicles market is shown in Fig. 1.

4.1.5 EV Incentive: From the survey, a moderate impact of EV incentives is observed. Governments around the world are offering various incentives to encourage consumers to purchase EVs. In India, the reduced GST rate on EVs is one such incentive that makes EVs more affordable and accessible to consumers. In 2022, the GST on EVs was reduced from 12 to 5%, and the GST on charging stations was reduced from 18 to 5%. In 2018, the GST on Li-ion batteries was reduced from 28 to 18%. This boost for EVs is significant as batteries account for nearly 50% of the cost of an EV [24, 25].

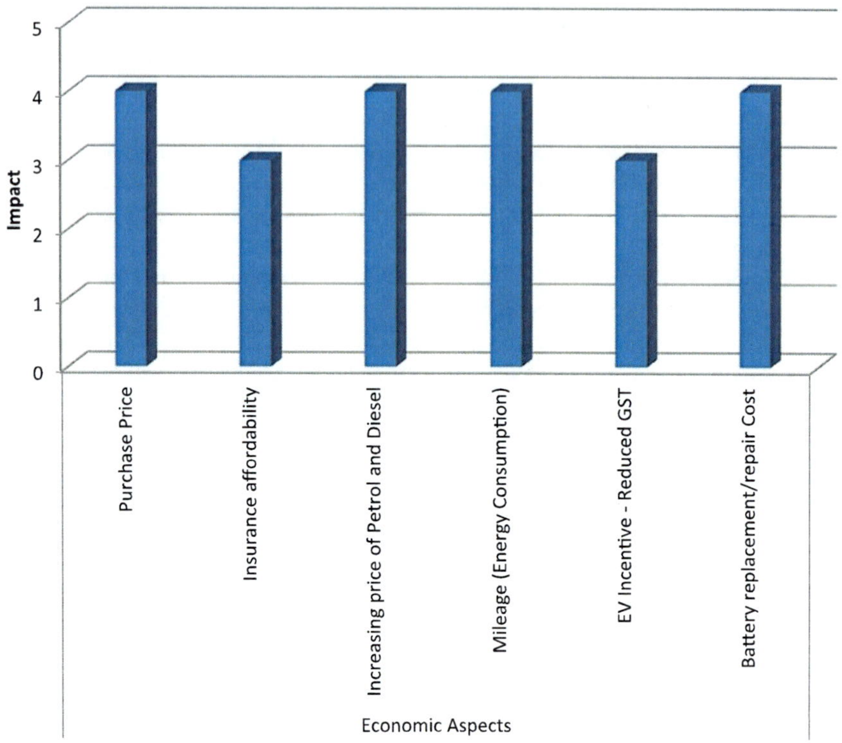

Fig. 1 Impact of economic aspects on electric vehicles market

4.1.6 Battery Replacement/Repair Cost: From the survey, a strong impact of Battery replacement/repair is observed. The cost of replacing or repairing an EV's battery is a critical factor that affects consumer perceptions and expectations. The replacement or repair of an EV's battery can be expensive, and consumers want to know the cost of these services and the warranty offered on the battery. The cost and frequency of battery replacement is a major factor that affects the purchasing decision for EVs [14].

4.2 Maintenance

4.2.1 Cost of Maintenance: From the survey, a moderate impact of cost of maintenance is observed. The cost of maintaining an EV is another factor that affects consumer perceptions and expectations. EVs are relatively new, and many consumers are uncertain about the cost of maintaining these vehicles. However, as EVs become more popular, the cost of maintenance is expected to decrease. The impact of maintenance aspects on electric vehicles market is shown in Fig. 2.

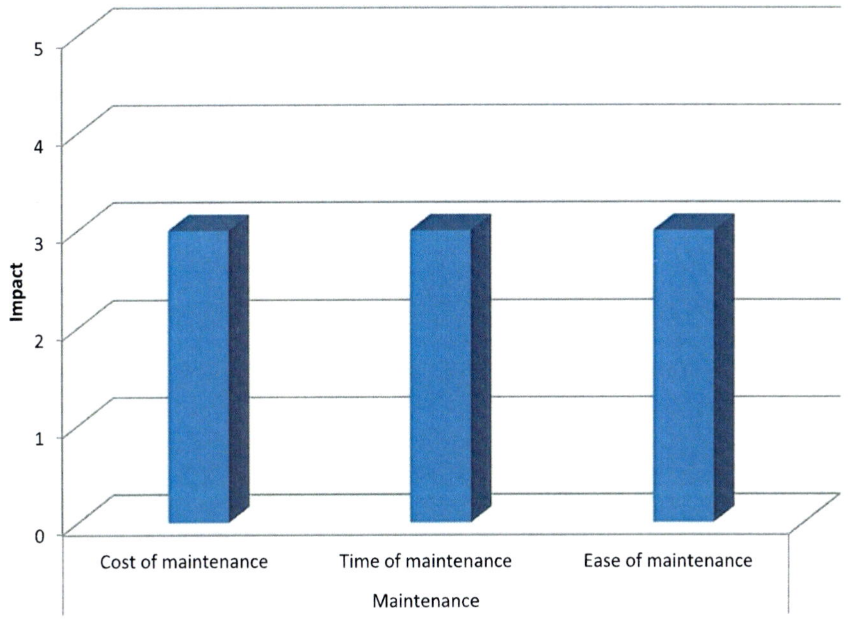

Fig. 2 Impact of maintenance aspects on electric vehicles market

4.2.2 Time of Maintenance: From the survey, a moderate impact of time required for maintenance is observed. The time required for maintenance is also an important factor that affects consumer perceptions and expectations. Consumers want to know how long it will take to perform maintenance on an EV, and whether the vehicle will be unavailable during this time.

4.2.3 Ease of Maintenance: From the survey, a moderate impact of ease of maintenance is observed. Consumers want to know how easy it is to perform maintenance on an EV. They want to know if they can perform routine maintenance themselves, or if they need to take the vehicle to a specialist. Due to the limited number of parts in an EV, maintenance costs and time are reduced unless there is a major service [17].

4.3 Battery and Charging Aspects

4.3.1 Charging Time: From the survey, a moderate impact of charging time is observed. The time required to charge an EV is another important factor that affects consumer perceptions and expectations. Consumers want to know how long it takes to charge an EV, and whether the charging process is convenient and accessible.

4.3.2 Availability of Charging Infrastructure: From the survey, a low impact of availability of charging infrastructure is observed. The availability of charging infrastructure is also a crucial factor that affects consumer perceptions and expectations. Consumers want to know if there are charging stations available in their area, and if they are convenient and accessible. Insufficient charging stations as well as the time required for charging EVs is a major hindrance to EV purchase decisions [14]. EV batteries have an estimated life of approximately 15–20 years [18]. Usually, EV battery manufacturers provide a good warranty of 8 years or a drive limit of 160,000 km [19]. An EV has an average capacity of 40 kWh, whereas top models reach up to100 kWh [20]. EV battery performance is severely affected above 30 °C and the battery is damaged beyond 40 °C [21]. This seems due to the presence of hybrid cars, better battery capacities, and the present limitations of EVs in urban areas [29]. Further, alternatives such as home-based charging stations and public charging stations located in convenient locations such as shopping centres, and parking garages may reduce the problem further.

4.3.3 Battery Life: From the survey, a moderate impact of Battery Life is observed. Consumers have high expectations for the battery life of electric vehicles. They expect their EVs to have a long-lasting battery that will keep the vehicle powered for an extended period. The average consumer wants an electric vehicle that can drive for at least 200 miles on a single charge.

4.3.4 Battery Warranty: From the survey, a moderate impact of Battery Warranty is observed. Consumers also expect a warranty for their electric vehicle's battery. They expect the battery to be covered for at least 5 years or 100,000 miles. This will give them peace of mind knowing that their battery will be covered in case of any defects or problems.

4.3.5 Battery Capacity: From the survey, a moderate to strong impact of Battery Capacity is observed. Consumers expect a battery capacity that is commensurate with the size of the vehicle. They expect a larger battery capacity for larger vehicles and vice versa. They also expect the battery to be easily replaceable if needed.

4.3.6 Battery Overheating Problem: From the survey, a moderate impact of the Battery Overheating problem is observed. Consumers are concerned about the safety of electric vehicles and are aware of the potential risks associated with battery overheating. They expect the manufacturers to have taken the necessary precautions to prevent such incidents and to provide a solution in case of an overheating problem. The impact of battery and charging aspects on the electric vehicles market is shown in Fig. 3.

4.4 EV Aspects

4.4.1 EV Warranty: From the survey, a moderate impact of EV warranty is observed. Consumers expect the electric vehicle warranty to be comprehensive and

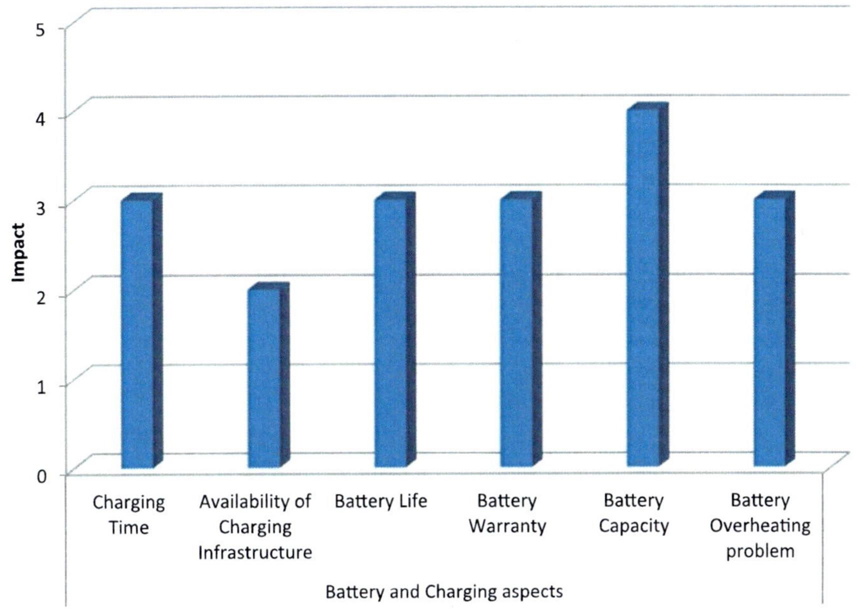

Fig. 3 Impact of battery and charging aspects on electric vehicles market

include coverage for the battery, charging system, and other electrical components. They also expect the warranty to be transferable to a new owner in case they sell their vehicle.

4.4.2 After Sales Service and Support: From the survey, a moderate impact of 'after sales service and support is observed. Consumers expect high-quality after-sales service and support for their electric vehicles. They expect the manufacturers to have a strong network of service centres and trained technicians to handle any problems or maintenance needs.

4.4.3 Vehicle Weight: From the survey, a moderate impact of vehicle weight is observed. Consumers expect electric vehicles to be light and have a high power-to-weight ratio. This will result in improved handling and better fuel efficiency. EVs equipped with 20 kWh Li batteries can weigh up to 200 kg [22].

4.4.4 Driving Range: From the survey, a moderate impact of driving range is observed. Consumers expect electric vehicles to have a driving range of at least 200 miles on a single charge. This will provide them with the peace of mind that they can travel long distances without having to worry about running out of power. Hybrids, Plug-In Hybrids, Extended-Range Electric Vehicles, Fuel Cell EVs, and Battery Electric Vehicles can travel approximately 25, 50, 390, 650, and 250 km on a single charge, respectively [30–35]. On a full charge, the average driving range of an EV is between 200 and 350 km.

However, top models like Tesla and Lucid Air have been reported to reach up to 500 km [23].

4.4.5 Aesthetics: From the survey, a moderate impact of aesthetics is observed. Consumers expect electric vehicles to be stylish and modern in design. They expect the vehicles to have a sleek and attractive look that will make them stand out on the road.

4.4.6 Software Update: From the survey, a moderate to strong impact of software updates is observed. Consumers expect electric vehicles to have the latest software updates that will improve the vehicle's performance and efficiency. They also expect the manufacturers to provide regular software updates to keep their vehicles up to date. The impact of EV aspects on the electric vehicles market is shown in Fig. 4.

4.4.7 Hardware Update: From the survey, a moderate impact of Hardware Update is observed. Consumers expect the manufacturers to provide hardware updates to their electric vehicles as well. This will ensure that the vehicle stays up to date with the latest technology and advancements.

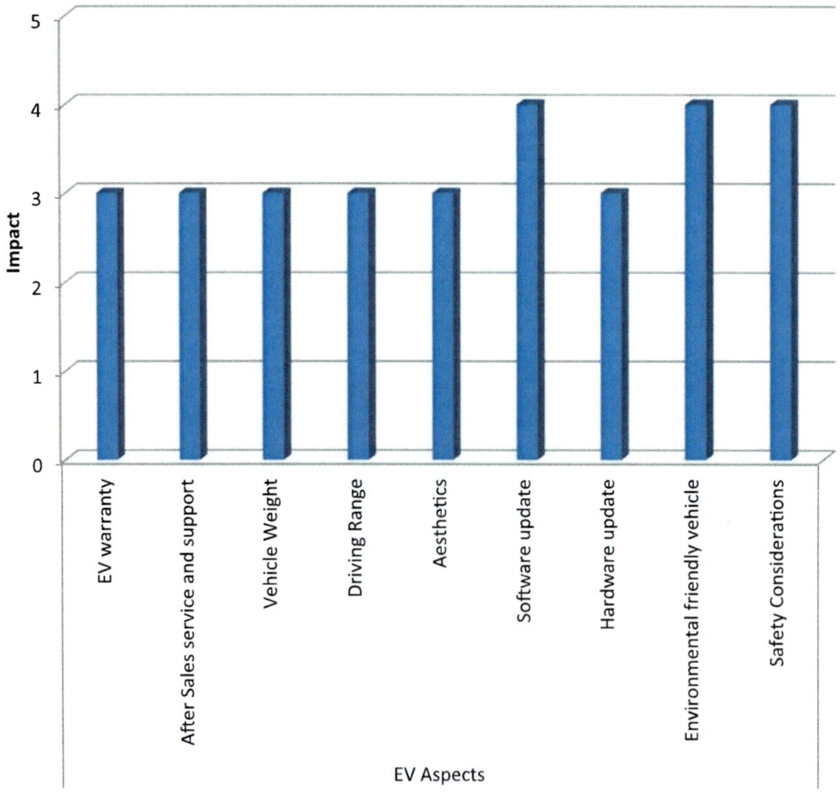

Fig. 4 Impact of EV aspects on electric vehicles market

4.4.8 Environmental Friendly Vehicle: From the survey, a strong impact of environmental friendliness of EV is observed. Consumers expect electric vehicles to be environmentally friendly. They expect the vehicles to have low emissions and to be powered by renewable energy sources. The use of EVs offers numerous benefits, including reduced emissions of harmful gases, lower noise pollution, reduced health hazards, a smaller carbon footprint, lower running and maintenance costs, and fuel conservation [15, 24, 25, 35].

4.4.9 Safety Considerations: From the survey, a strong impact of safety considerations is observed. Consumers expect electric vehicles to be safe to drive. They expect the vehicles to have advanced safety features such as airbags, anti-lock brakes, and stability control. They also expect the vehicles to be crash-tested and meet the latest safety standards.

4.5 EV Performance Aspects

4.5.1 Power: From the survey, a moderate to strong impact of power is observed. Consumers expect electric vehicles to have high power and instant torque delivery. This will provide them with a smooth and enjoyable driving experience. Power of the vehicle is good as compared to conventional IC engine-based vehicle. However, consumers have rated acceleration, torque and top speed to be moderate. IC Engine based vehicles need time to build rpm to produce high torque whereas EVs produce high torque instantly at all speeds [16]. Due to this, the EVs seem to be quite powerful to the driver. EV acceleration is better than those of IC engine-based vehicle in EVs with high-capacity batteries only [16]. The top speeds of EV vehicles are good and reaches around 80% of that of IC engine-based vehicle [26].

4.5.2 Acceleration: From the survey, a moderate impact of acceleration is observed. Consumers expect electric vehicles to have quick acceleration. They expect the vehicles to be able to go from 0 to 60 miles per hour in less than 7 s.

4.5.3 Torque: From the survey, a moderate impact of torque is observed. Consumers expect electric vehicles to have high torque. This will provide them with excellent acceleration and a smooth driving experience.

4.5.4 Top Speed: From the survey, a moderate impact of top speed is observed. One of the most critical parameters for consumer perceptions and expectations of electric vehicles is their top speed. Electric vehicles are often criticized for their slow speed compared to traditional gasoline-powered cars. However, this perception is changing with advancements in technology that are increasing the top speed of electric vehicles. Consumers today expect electric vehicles to have a top speed of at least 80–90 mph, with high-end models capable of reaching 120 mph. The impact of EV performance aspects on the electric vehicles market is shown in Fig. 5.

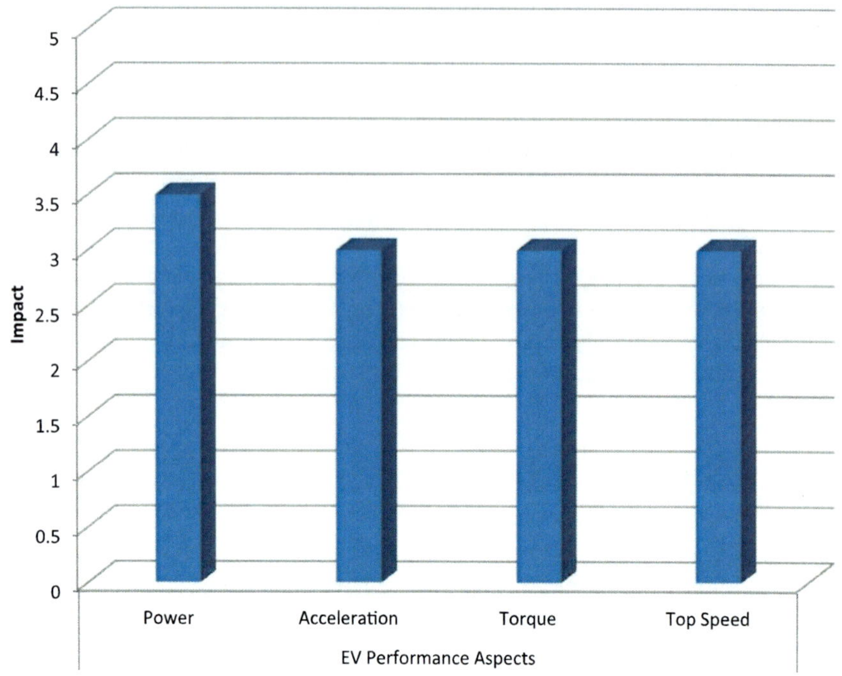

Fig. 5 Impact of EV performance aspects on electric vehicles market

5 Additional Aspects

Additional aspects that are not considered presently but are likely to gain importance in the future are the EV resale price and the end of life disposal.

5.1 Resale Price

Another critical factor that affects consumer perceptions and expectations of electric vehicles is their resale price. Consumers expect electric vehicles to have a high resale price due to their longer lifespan and higher reliability compared to gasoline-powered vehicles. However, the reality of the resale price is often different, with electric vehicles losing a significant portion of their value compared to traditional cars. Consumers expect the resale price of electric vehicles to be at least on par with traditional vehicles and for their depreciation to be minimal.

5.2 *End-of-Life Disposal*

Consumers are also concerned about the end-of-life disposal of electric vehicles, including their batteries and other components. Consumers expect electric vehicle manufacturers to have a clear plan in place for the disposal of their products, ensuring that the environment is not affected. This is becoming increasingly important as consumers become more environmentally conscious and look for sustainable solutions.

The summary of the dominant response for the impact of sub-factors on the electric vehicles market is shown in Table 1.

6 Conclusions

The following aspects can be concluded from the research:

1. The purchase price and the cost of insuring EVs are currently higher, but expected to decrease as the market for EVs grows.
2. The increasing price of gasoline and diesel is making EVs an attractive alternative as they are cheaper to operate and maintain.
3. The driving range of EVs is increasing with advancements in battery technology and the cost of recharging is becoming more affordable.
4. Governments are offering various incentives to encourage the adoption of EVs, including reduced GST rates in India.
5. The cost of maintaining EVs is expected to decrease as they become more popular, and maintenance is relatively easy with a limited number of parts.
6. The time required to charge an EV and the availability of charging infrastructure are also important factors that affect consumer perceptions and expectations.
7. EVs offer numerous benefits such as zero emissions, low maintenance, and a smoother and quieter driving experience.
8. The impact of purchase price, increasing price of petrol and diesel, and mileage were found to have moderate to very strong impacts on the EV market. Battery capacity, software update, and power were found to have a moderate to strong impact.
9. Battery replacement/repair, environmental friendliness, and safety considerations were found to have a strong impact.

7 Future Scope

The research can further be extended to the ranking of the sub-factors based on qualitative and quantitative analysis of their impact on the electric vehicles market.

Table 1 Summary of dominant response to the impact of sub-factors on electric vehicles market

Sr. No	Factor	Sub-factor	Dominant response to impact of sub-factor on electric vehicles market
1	Economic aspects	Purchase price	Moderate to very strong
		Insurance affordability	Moderate
		Increasing price of petrol and diesel	Moderate to very strong
		Mileage (energy consumption)	Moderate to very strong
		EV incentive—reduced GST	Moderate
		Battery replacement/ repair cost	Strong
2	Maintenance	Cost of maintenance	Moderate
		Time of maintenance	Moderate
		Ease of maintenance	Moderate
3	Battery and charging aspects	Charging time	Moderate
		Availability of charging Infrastructure	Low
		Battery life	Moderate
		Battery warranty	Moderate
		Battery capacity	Moderate to strong
		Battery overheating problem	Moderate
4	EV aspects	EV warranty	Moderate
		After sales service and support	Moderate
		Vehicle weight	Moderate
		Driving range	Moderate
		Aesthetics	Moderate
		Software update	Moderate to strong
		Hardware update	Moderate
		Environmental friendly vehicle	Strong
		Safety considerations	Strong
5	EV performance aspects	Power	Moderate to strong
		Acceleration	Moderate
		Torque	Moderate
		Top speed	Moderate

References

1. Albatayneh A, Assaf MN, Alterman D, Jaradat M (2020) Comparison of the overall energy efficiency for internal combustion engine vehicles and electric vehicles. Environ Clim Technol 24:669–680. https://doi.org/10.2478/rtuect-2020-0041
2. Chan CC (2007) The state of the art of electric, hybrid, and fuel cell vehicles. Proc IEEE 95:704–718. https://doi.org/10.1109/JPROC.2007.892489
3. Supercharger. https://www.tesla.com/supercharger
4. California Bill Would Ban Gas Cars By 2040. https://www.newsweek.com/california-ban-gas-powered-cars-2040-740584
5. New Dutch government's plans for the coming years. https://nltimes.nl/2017/10/10/new-dutch-governments-plans-coming-years
6. Paris to ban diesel cars from 2024, all internal combustion vehicles from 2030|DriveMag Cars (2017) https://drivemag.com/news/paris-to-ban-diesel-cars-from-2024-all-internal-combustion-vehicles-from-2030/
7. Connolly K (2018) German court rules cities can ban diesel cars to tackle pollution. https://www.theguardian.com/environment/2018/feb/27/german-court-rules-cities-can-ban-diesel-cars-to-tackle-pollution
8. Dow J Rome latest city to announce car ban, will ban diesel cars from historical center starting 2024. https://electrek.co/2018/02/28/rome-bans-diesel-cars-2024/
9. Petroff A These countries want to ban gas and diesel cars. https://money.cnn.com/2017/09/11/autos/countries-banning-diesel-gas-cars/index.html
10. Solomon S Israel aims to eliminate use of coal, gasoline and diesel by 2030, https://www.timesofisrael.com/israel-aims-to-eliminate-use-of-coal-gasoline-and-diesel-by-2030/
11. Why electric cars insurance rates are expensive? | Tata AIG Blog. https://www.tataaig.com/knowledge-center/car-insurance/why-electric-cars-expensive-to-insure
12. Desk HA Oil price hike to fuel Indian EV market growing to 30% by 2030: Study. https://auto.hindustantimes.com/auto/electric-vehicles/oil-price-hike-to-fuel-ev-sales-growth-in-india-claims-study-41647961518638.html
13. Singh R (2021) Electric car vs petrol car: cost of running, maintenance, and everything else you need to know|91mobiles.com. https://www.91mobiles.com/hub/electric-car-vs-petrol-car/
14. Livemint: Cost and frequency of battery replacement a major barrier to buying an EV: report. https://www.livemint.com/news/india/cost-and-frequency-of-battery-replacement-a-major-barrier-to-buying-an-ev-report-11670246973742.html
15. Benefits of electric vehicles. https://e-amrit.niti.gov.in/benefits-of-electric-vehicles
16. Bharadwaj P Why do electric cars accelerate faster than internal combustion cars? https://www.drivespark.com/off-beat/why-do-electric-cars-accelerate-faster-than-internal-combustion-cars-explained-035208.html
17. EV maintenance costs vs petrol/diesel cars: are electric cars cheaper to own? (2022) https://timesofindia.indiatimes.com/auto/electric-cars/ev-maintenance-costs-vs-petrol/-diesel-cars-are-electric-cars-cheaper-to-own/articleshow/94322087.cms
18. Barlingen W. van: How long do electric car batteries last? | EVBox, https://blog.evbox.com/ev-battery-longevity
19. BU-1003a: Battery aging in an electric vehicle (EV). https://batteryuniversity.com/article/bu-1003a-battery-aging-in-an-electric-vehicle-ev
20. Electric vehicles|Battery|Capacity and Lifespan. https://www.eonenergy.com/electric-vehicle-charging/costs-and-benefits/battery-capacity-and-lifespan.html
21. Shepard J EV battery thermal management challenges. https://www.batterypowertips.com/ev-battery-thermal-management-challenges-faq/
22. Berjoza D, Jurgena I (2017) Effects of change in the weight of electric vehicles on their performance characteristics. Agron Res 15:952–963
23. Lambert F Elon Musk: Tesla could have made a 600-mile-range electric car a year ago. https://electrek.co/2022/03/02/elon-musk-tesla-could-have-made-600-mile-range-electric-car/

24. Alimujiang A, Jiang P (2020) Synergy and co-benefits of reducing CO_2 and air pollutant emissions by promoting electric vehicles—a case of Shanghai. Energy Sustain Dev 55:181–189. https://doi.org/10.1016/j.esd.2020.02.005
25. Malmgren I (2016) Quantifying the societal benefits of electric vehicles. World Electr Veh J 8:996–1007. https://doi.org/10.3390/wevj8040996
26. Are electric cars faster than gas ones?|GetJerry.com https://getjerry.com/electric-vehicles/are-electric-cars-faster-gas-ones
27. Electric vehicles: literature review of technology costs and carbon emissions. https://theicct.org/publication/electric-vehicles-literature-review-of-technology-costs-and-carbon-emissions/
28. Notter DA, Gauch M, Widmer R, Wäger P, Stamp A, Zah R, Althaus H-J (2010) Contribution of Li-Ion batteries to the environmental impact of electric vehicles. Environ Sci Technol 44:6550–6556. https://doi.org/10.1021/es903729a
29. Maherchandani JK, Joshi RR, Tirole R, Swami RK, Ganthia BP (2022) Performance comparison analysis of energy management strategies for hybrid electric vehicles. In: Kumar S, Singh B, and Singh AK (eds) Recent advances in power electronics and drives. Springer Nature, Singapore, pp 245–254
30. 2014 Toyota Prius PHV: To plug in or not to plug in? https://www.guideautoweb.com/en/articles/21152/
31. 2019 BMW i3, i3 REx, i3s & amp; i3s REx: Full Specs. https://insideevs.com/news/339970/2019-bmw-i3-i3-rex-i3s-amp-i3s-rex-full-specs/
32. 2023 Mitsubishi Outlander PHEV Plug-In Hybrid SUV|Mitsubishi Motors. https://www.mitsubishicars.com/cars-and-suvs/outlander-phev
33. NEXO press kit. https://www.hyundai.news/eu/models/electrified/nexo/press-kit.html
34. Nissan Reveals LEAF e-Plus: 62 kWh Battery, 226-Mile Range. https://insideevs.com/news/341958/nissan-reveals-leaf-e-plus-62-kwh-battery-226-mile-range/
35. Gupta S, Poonia S, Varshney T, Swami RK, Shrivastava A (2022) Design and implementation of the electric bicycle with efficient controller. In: Tripathi A, Soni, A, Shrivastava A, Swarnkar A, and Sahariya J (eds) Intelligent computing techniques for smart energy systems, Springer Nature, Singapore, pp 541–552

Service Quality in Health Care: A Bibliometric Analysis

Mohit Datt, Ajay Gupta, and Sushendra Kumar Misra

Abstract A bibliometric study of service quality in health care was performed due to its growing importance and to examine service quality trends in existing healthcare literature. The research articles related to the quality of healthcare services were assessed by using the Bibliometrix software. Using the Scopus database, 1202 documents were retrieved and analyzed in the present study. The co-occurrence analysis reported that China and the United States were leading countries in this domain. Also, year-wise trends, most valuable sources, most frequent keywords, and leading authors were identified. This paper recommends valuable insights for policymakers, practitioners, and researchers to learn the potential of service quality in healthcare. This article will influence future needs and trends in healthcare service quality.

Keywords Service quality · Healthcare · Bibliometric analysis

1 Introduction

In the current era, the focus on quality has become the key to success for the organizations. The concept of quality is equally important for the service sector, but initially it originated in the manufacturing sector. Quality is now acknowledged as a crucial tactical tool for achieving effectiveness in operations and enhancing organizations' performance. Similarly with the increasing competition in the healthcare sector, the customers are demanding high quality of care during the services and consumers' opinions help to improve healthcare quality.

According to Endewshaw [1], every country and healthcare service organization should have its own quality measurement framework. Hence in the literature for

M. Datt (✉) · A. Gupta
Department of Industrial and Production Engineering, Dr B R Ambedkar National Institute of Technology, Jalandhar 144027, India
e-mail: mohitdatt1991@gmail.com

S. K. Misra
I.K. Gujral Punjab Technical University, Jalandhar, India

A. Sachdeva et al. (eds.), *Recent Advances in Operations Management and Optimization*, Lecture Notes in Mechanical Engineering,
https://doi.org/10.1007/978-981-99-7445-0_21

measuring the service quality (SQ) of healthcare many authors developed, tested, and implemented the SQ models namely, HospitalQual, HEALTHQUAL, PubHosQual according to their demographic environments [2–4].

Bibliometric analysis is an established and efficient tool to study enormous scientific datasets. This highlights emerging trends about a particular topic because it analyses keywords, the number of citations, and collaboration patterns between authors and countries. It suggests that the academicians/practitioners to overview the contributions of other scholars about any required domain. Bibliometric analysis is performed in different disciplines and contexts including marketing [5], management science [6, 7], tourism [8], environmental concerns [9], and manufacturing [10].

Erstwhile, many researchers reported the importance of SQ in healthcare and measured the SQ in different healthcare environments of healthcare settings. Therefore, this article aims to address the following research questions:

RQ1. Describe the trend of applications of SQ in health care?
RQ2. Describe the key findings of SQ in healthcare from the existing literature?

2 Methodology

This research paper investigates the state-of-the-art on SQ in health care. The main objective of this review is to identify the emerging themes in the domain of SQ in health care. We perform this study by using the three-phase systematic process with investigation guidelines where phase–I for planning the review, phase–II for conducting the review, and phase–III for reporting and dissemination [11].

Phase–I: Initially research articles were downloaded from the Scopus database for analysis. Because of its largest abstract and citation database of peer-reviewed literature [12]. Data was retrieved from the database using the search strategy ((“Service Quality”) AND ((“Healthcare”) OR (“Hospital”) OR (“Clinic”))) and searched within article title, abstract, and keywords. The period covered for this study is from January 2013 to Mid-February 2023.

Phase–II: Using the above setting criteria, a total of 1999 articles were retrieved from the database. Then, we removed 797 records on the basis of inclusion criteria. We considered only journal, conference proceedings, and review articles included for further study of data. In addition, this research shortlisted papers published in English language only. Finally, 1202 articles were retrieved for bibliometric analysis.

Phase–III: During this phase, the articles were analyzed with the Bibliometrix software for year-wise trends, most valuable sources, most frequent keywords, leading authors, and co-occurrence network (Fig. 1).

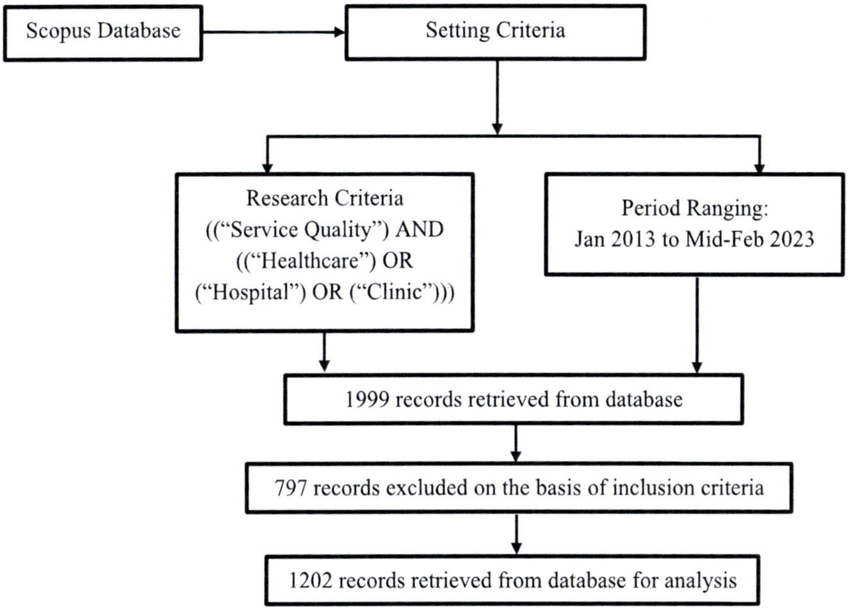

Fig. 1 Research methodology

3 Results and Discussions

This section discusses the results of the study based on the bibliometric analysis and to understand the trends in year-wise publications, authors’ contribution, and keywords used. All the observations are discussed in this section below.

3.1 Primary Information About Data

The main information about the data is represented in Fig. 2. In this data total of 4359 authors and more than 90 percent of articles in which more than a single author contributed, which means the collaboration between the academicians in this domain. References of more than 57,000 indicate the broad concept of service quality in healthcare.

3.2 Year-Wise Publication

Figure 3 shows the trends of published articles for the past decade from 2013 to 2022. There was regular significant growth in the number of publications in the domain of

Timespan 2013: Mid-Feb 2023	**Sources** 585	**Documents** 1202
Authors 4359	**Single-authored docs** 94	**Document average age** 4.02
Average citations per doc 11.19	**References** 57460	**International co-authorship** 26.54%
Co-authors per doc 4.04	**Author's keywords (DE)** 3348	**Articles 1121, conference paper 4, review 77**

Fig. 2 Primary data for research

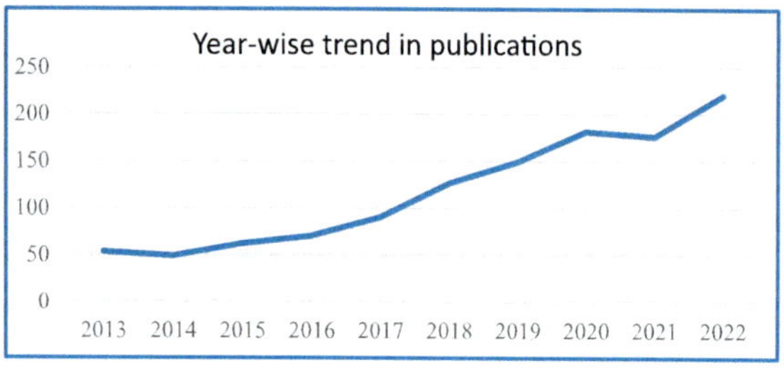

Fig. 3 Evolution of year-wise publications

SQ in healthcare. Here it means the interest of researchers is constantly increasing in this domain. This trend represents the positive annual growth rate which motivates new researchers to explore this area for the betterment of SQ for customers in health care.

3.3 *Most Valuable Sources*

Table 1 shows the involvement of different sources in this domain. The R package discovered a total of 585 sources and out of these here we represent the ten most relevant sources in tabular form. "International Journal of Environment Research and Public Health" was the leading journal followed by "International Journal of Health Care Quality Assurance" and "BMC Health Service Research". All these reputed journals publish only qualitative work that demonstrates relevance for social and healthcare settings.

Table 1 Top ten sources in domain of SQ in health care

Name of journal	Total publications
International Journal of Environmental Research and Public Health	102
International Journal of Health Care Quality Assurance	68
BMC Health Services Research	37
International Journal of Healthcare Management	25
BMJ Open	15
Total Quality Management and Business Excellence	15
Frontiers in Public Health	13
International Journal of Pharmaceutical and Healthcare Marketing	13
Social Science and Medicine	12
TQM Journal	12

3.4 *Three Field Point Representation*

Figure 4 shows three plots between keywords, authors, and countries, which helps to determine countries who are leading in this domain and international co-authorships of authors. It also highlights the linking of crucial keywords with different authors.

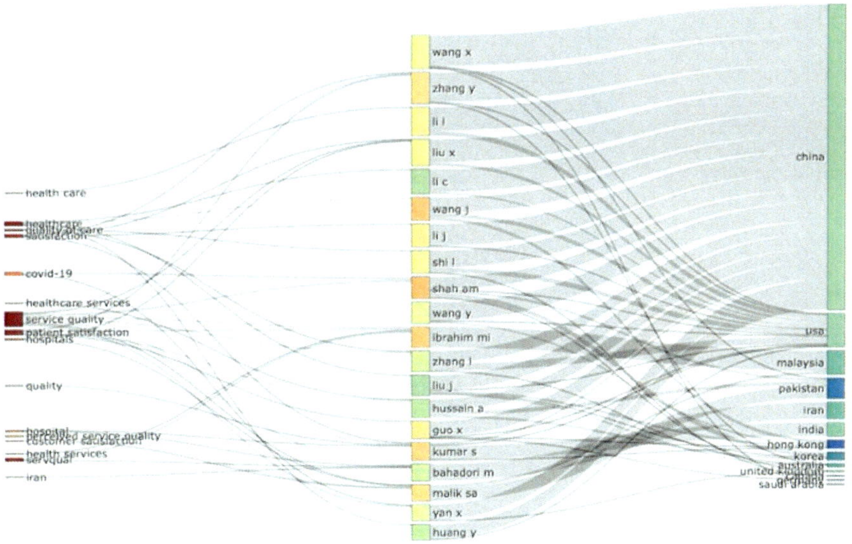

Fig. 4 Three field point graph between keywords, authors, and countries

3.5 Most Frequent Words

We used R studio software to draw a word cloud of keywords for this study. The most common keyword for this study was human followed by female, humans, male, adult, article, patient satisfaction, healthcare care quality, middle aged, and healthcare. Here we discussed the most common ten keywords in the domain of SQ in healthcare (refer to Table 2). Figure 5 demonstrates the pictorial view of all the other words such as service quality, primary healthcare, health services, quality of healthcare, quality of service, etc. We conclude SQ in health care directly targets the population and it is most important for human beings.

Table 2 Mostly used words

Words	Occurrences
Human	643
Female	558
Humans	484
Male	483
Adult	462
Article	416
Patient satisfaction	335
Health care quality	287
Middle aged	242
Health care	220

Fig. 5 Pictorial view of most commonly used words

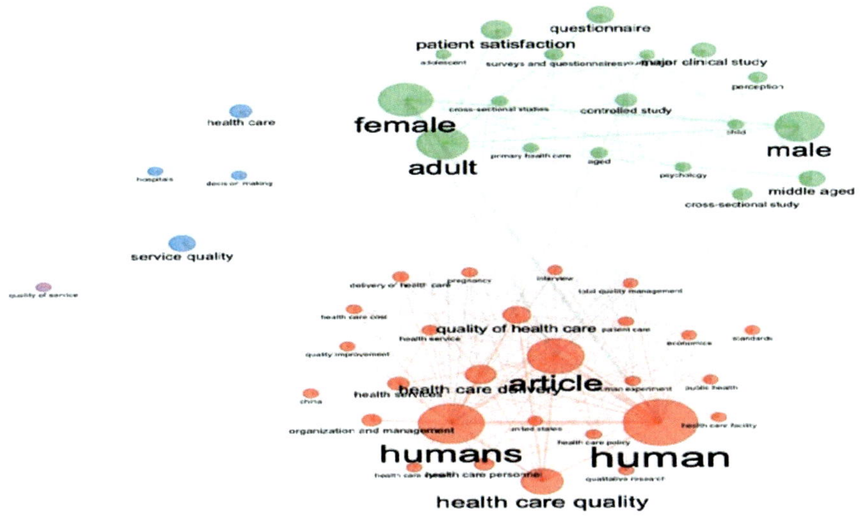

Fig. 6 Graphically representation of co-occurrence network

3.6 *Co-occurrence Network*

This study analyses 158 co-occurrence networks. The graphical representation includes 0.7 opacity, 56 labels, and eight label sizes. This graph shows the clusters of all the keywords used in the domain of SQ in healthcare. Network allows us to visualize the interferences regarding relationships between the words in this domain (Fig. 6).

3.7 *Average Citation per Year*

See Table 3.

3.8 *Leading Authors*

The R studio identified 4359 authors and on the basis of h-index, g-index, and m-index we selected the top ten authors in the domain of SQ in health care. Malik, S. A., is the most contributing author followed by Guo, X., and Zhang, L (Table 4).

Table 3 Average citations per year

Year	N	MeanTCperArt	MeanTCperYear	CitableYears
2013	53	25.30	2.53	10
2014	49	25.86	2.87	9
2015	62	22.52	2.81	8
2016	70	16.07	2.30	7
2017	90	18.60	3.10	6
2018	126	15.77	3.15	5
2019	149	13.99	3.50	4
2020	181	8.11	2.70	3
2021	176	4.49	2.24	2
2022	220	1.35	1.35	1
2023	26	0.81		0

Table 4 List of leading 10 authors

Element	h_index	g_index	m_index	TC	NP	PY_start
Malik SA	6	6	0.6	306	6	2014
Guo X	5	5	0.5	231	5	2014
Zhang L	5	5	0.556	81	5	2015
Bahadori M	4	4	0.4	46	4	2014
Hussain A	4	4	0.8	127	4	2019
Ravangard R	4	4	0.4	46	4	2014
Shah AM	4	7	1	57	7	2020
Wu T	4	4	0.444	216	4	2015
Zhang Y	4	5	0.667	46	5	2018
Anabila P	3	3	0.6	32	3	2019

4 Conclusions

The aim of this research was to conduct a bibliometric analysis of SQ in the healthcare sector. After bibliometric analysis, 1202 records were retrieved by using the Scopus database from the year 2013 to Mid-February 2023. Study reported annual growth of publications and co-occurrence analysis was performed for SQ in healthcare. SQ analysis improves the performance of healthcare settings and guides the administrators in highly demanding healthcare quality in the healthcare context.

Here in this study, we used only the Scopus database and hence database barrier is one of the limitation and in future analysis can be performed by using other databases like Google Scholar, Web of Science. Many models proposed by researchers in the domain of SQ in healthcare and in future it is suggested to perform bibliometric analysis with keywords such as HEALTHQUAL, PubHosQual, and HospitalQual.

Also, this study suggests performing future research in the combination of SQ with brand image, patient loyalty, and behavioral intention.

References

1. Endeshaw B (2021) Healthcare service quality-measurement models: a review. J Health Res 35(2):106–117
2. Itumalla R, Acharyulu GVRK, Shekhar BR (2014) Development of hospitalqual: a service quality scale for measuring in-patient services in hospital. Oper Supply Chain Manag Int J 7(2):54–63
3. Lee D (2017) HEALTHQUAL: a multi-item scale for assessing healthcare service quality. Serv Bus 11:491–516
4. Aagja JP, Garg R (2010) Measuring perceived service quality for public hospitals (PubHosQual) in the Indian context. Int J Pharm Healthc Market 4(1):60–83
5. Martínez-López FJ, Merigó JM, Valenzuela-Fernández L, Nicolás C (2018) Fifty years of the European Journal of Marketing: a bibliometric analysis. Eur J Mark 52(1/2):439–468
6. Merigó JM, Yang JB (2017) A bibliometric analysis of operations research and management science. Omega 73:37–48
7. Ikeziri LM, Souza FBD, Gupta MC, de Camargo Fiorini P (2019) Theory of constraints: review and bibliometric analysis. Int J Prod Res 57(15–16):5068–5102
8. Park J, Jeong E (2019) Service quality in tourism: a systematic literature review and keyword network analysis. Sustainability 11(13):3665
9. Zhang X, Estoque RC, Xie H, Murayama Y, Ranagalage M (2019) Bibliometric analysis of highly cited articles on ecosystem services. PLoS ONE 14(2):e0210707
10. Gundu K, Jamwal A, Yadav A, Agrawal R, Jain JK, Kumar S (2022) Circular economy and sustainable manufacturing: a bibliometric based review. In Recent Advances in Industrial Production: Select Proceedings of ICEM 2020:137–147
11. Tranfield D, Denyer D, Smart P (2003) Towards a methodology for developing evidence-informed management knowledge by means of systematic review. Br J Manag 14(3):207–222
12. Khudzari JM, Kurian J, Tartakovsky B, Raghavan GSV (2018) Bibliometric analysis of global research trends on microbial fuel cells using Scopus database. Biochem Eng J 136:51–60

A Vision-Based Hand Gesture Recognition System: Development and Modelling

Devanshu Chaudhary, Bhargav Aggarwal, Yatharth, Agrim Gupta, and Nitin Dixit

Abstract Gesture recognition systems are gaining popularity nowadays because they can interface with machines via human–computer interaction. The development of such systems helps to develop a more natural communication between man and machine. These setups made humans directly instruct the machine as per their needs. A vision-based gesture recognition setup has been developed and modelled in the present study. The gesture recognition system has been developed by a Linux processing device with a camera and an output control device. MediaPipe Hands model has been used for gesture recognition. Programming for the used model has been done using Python programming language. The Arduino UNO has also been programmed to convert the gestures into desired outputs. The developed vision-based gesture recognition system was found to work effectively for human hand gestures.

Keywords Gesture · Linux · MediaPipe · Python · Arduino

Nomenclature

OS	Operating System
IDE	Integrated Development Environment
ML	Machine Learning

D. Chaudhary · B. Aggarwal · Yatharth · A. Gupta · N. Dixit (✉)
Mechanical Engineering Department, Punjab Engineering College (Deemed to be University), Chandigarh 160012, India
e-mail: nitindixit@pec.edu.in

A. Sachdeva et al. (eds.), *Recent Advances in Operations Management and Optimization*, Lecture Notes in Mechanical Engineering,
https://doi.org/10.1007/978-981-99-7445-0_22

1 Introduction

Hand gesture detection systems have sparked considerable interest in the recent past because of their many applications and ability to interface with machines conveniently via human-to-computer interaction. Gesture recognition is a critical component of human–computer interaction. Computer vision-based gesture recognition systems have made humans convey their thoughts and instructions to robots [1]. The primary objective behind developing gesture recognition systems is to establish a natural communication between humans and computers to transmit relevant information [2]. However, sometimes it becomes difficult to generate gestures that can be recognized and interpreted well by the computer [3]. Techniques such as vision-based and glove/sensor-based approaches have been used to develop hand gesture recognition setups. In a vision-based system, the camera is used as an input device, whereas in a glove-based system, an artificial hand glove provides input to the gesture recognition system. Gloves offer a precise measure of hand attitude and movement but are pretty costly [4].

Various gesture recognition setups have been developed in the recent past. Damdoo et al. [5] developed an adaptive hand gesture recognition setup and reported that light conditions directly affect the recognition system's performance. The gesture recognition system shows better results in good light. A Hidden Markov Model-based gesture recognition setup was developed by Grobel and Assan [6] to identify isolated gestures from video. Yuanxin et al. [7] used a variational parameter model of image motion to recognize gestures. Ling et al. developed a self-learning space representation-based gesture system [8]. Wang et al. [9] proposed real-time gesture recognition using optical flow matching and the AdaBoost algorithm.

A vision-based gesture recognition setup has been developed in the present study. The gesture recognition system has been developed by a Linux processing device with a camera and an output control device. MediaPipe Hands model has been used for gesture recognition. Programming for the used model was done using Python programming language. The Arduino UNO has also been programmed to convert the gestures into desired outputs. The working of the developed setup has been tested using a 4 LED setup. The developed setup was responding effectively to different human hand gestures.

2 Gesture Recognition System

A gesture recognition system has been designed, developed and modelled to recognize human hand signals effectively.

2.1 Components of Gesture Recognition System

The gesture recognition system comprises a Linux processing device, MediaPipe, a camera, an Integrated Development Environment, an external control system, and an output device, as shown in Fig. 1. The details of these components are explained as follows.

2.1.1 Linux Processing Device

A processing device that can run any of the latest Linux Operating Systems (OS) is needed. Some feasible options for the operating system are Ubuntu LTS, Ubuntu Mint, Fedora OS, and Elementary OS. Linux OS has frequently available updates, and the latest dependency version can be found easily. Moreover, Linux OS can be run on processing devices with the standard configuration; hence, no higher-end hardware is needed.

2.1.2 MediaPipe

MediaPipe offers cross-platform, customizable Machine Learning (ML) solutions for live and streaming media. MediaPipe consists of several open-source ML frameworks such as face detection, face mesh, iris, hands, pose, holistic, hair segmentation, object

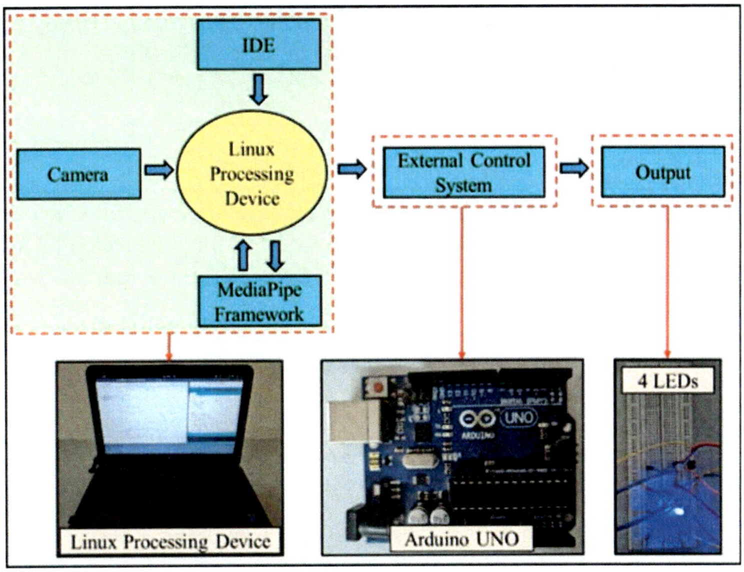

Fig. 1 Schematic and components of the gesture recognition system

detection, box tracking, instant motion tracking, objectron and KNIFT. Anyone can use each model under the Apache 2.0 license on Android, Python, C++, IOS and J.S. [10]. In the present gesture recognition system, the hands model has been modified and used as desired.

2.1.3 Integrated Development Environment (IDE)

IDE helps create a virtual environment on devices, making running programs easy. IDE is used to write the executing code of the MediaPipe framework in Python programming language.

2.1.4 Camera

A camera that supports 720p recording with a minimum of 20fps is usually needed to recognize the gestures. The camera is an input device for the MediaPipe framework.

2.1.5 External Control System

An external control system is used to control the outputs of the gesture recognition system. In the present work, an Arduino UNO microcontroller has been used to control the outcome. Arduino uses ATmega 328 chip responsible for executing the code programmed in it. Arduino receives signals from Linux processing devices and triggers different external circuit outputs. The Arduino is programmed using Arduino programming languages. The code for Arduino is presented in Fig. 2.

2.1.6 Output

In the present work, the working of the developed gesture recognition system was tested and verified using a 4-LED system. These LEDs are connected to Arduino, and different LEDs respond to a particular are triggered by the four other gestures of the gesture recognition model.

2.2 Gesture Recognition Model

Hand shape and motion perception are crucial in optimizing user experience in numerous applications. MediaPipe Hands model is used to develop a gesture recognition system for the present study. MediaPipe Hands is a solution for hand and finger tracking with high fidelity. It extracts 21 3D landmarks of the hand from the picture using ML and recognizes the hand gesture using palm detection and hand landmark

```
int a=2,b=3,c=4,d=5;
void setup() {
  // put your setup code here, to run once:
  Serial.begin(9600);
  pinMode(a,OUTPUT);
  pinMode(b,OUTPUT);
  pinMode(c,OUTPUT);
  pinMode(d,OUTPUT);
}
void loop() {
  // put your main code here, to run repeatedly:
  if(Serial.available() > 0){
    char fingers = Serial.read();
    //incomingByte=Serial.readStringUntil('\n');
    Serial.print(fingers);
    if(fingers=='1'){
      digitalWrite(a,HIGH);
      digitalWrite(b,LOW);
      digitalWrite(c,LOW);
      digitalWrite(d,LOW);
    }
    else if (fingers=='2'){
      digitalWrite(a,LOW);
      digitalWrite(b,HIGH);
      digitalWrite(c,LOW);
      digitalWrite(d,LOW);
    }
    else if (fingers=='3'){
      digitalWrite(a,LOW);
      digitalWrite(b,LOW);
      digitalWrite(c,HIGH);
      digitalWrite(d,LOW);
    }
    else if (fingers=='4'){
      digitalWrite(a,LOW);
      digitalWrite(b,LOW);
      digitalWrite(c,LOW);
      digitalWrite(d,HIGH);
    }
    else{
      digitalWrite(a,LOW);
      digitalWrite(b,LOW);
      digitalWrite(c,LOW);
      digitalWrite(d,LOW);
    }
  }
}
```

Fig. 2 Code for Arduino

detection. The palm detection model generates an oriented hand-bounding box from the image. The hand landmark model uses the palm detector's cropped picture region to detect 21 3D hand key points, as shown in Fig. 3. The final gesture thus obtained from the model is further processed by the external control system to generate the desired output.

The MediaPipe hand model has been coded using Python as presented in Fig. 4.

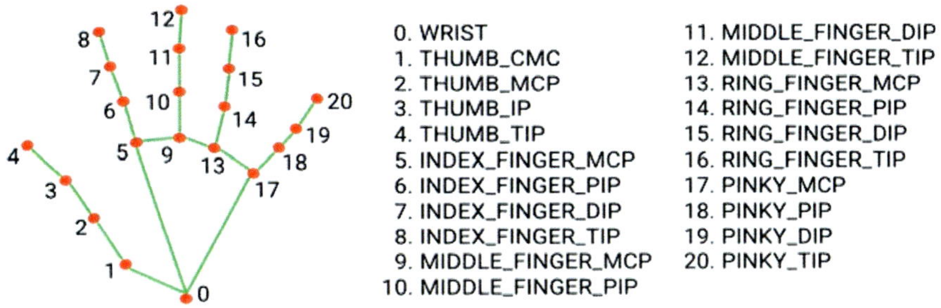

Fig. 3 Hand pose representation of MediaPipe hands [11]

```
import cv2
import mediapipe as mp
import time
class handDetector():
        def __init__
        (self, mode=False, maxHands=1, detectionCon=0.5, trackCon=0.5):
                self.mode = mode
                self.maxHands = maxHands
                self.detectionCon = detectionCon
                self.trackCon = trackCon
                self.mpHands = mp.solutions.hands
                self.hands = self.mpHands.
                Hands(self.mode,self.maxHands,self.detectionCon,self.trackCon)
                self.mpDraw = mp.solutions.drawing_utils
        def findHands(self, img, draw=True):
                imgRGB = cv2.cvtColor(img, cv2.COLOR_BGR2RGB)
                self.results = self.hands.process(imgRGB)
                # print(results.multi_hand_landmarks)
                if self.results.multi_hand_landmarks:
                for handLms in self.results.multi_hand_landmarks:
                if draw:
                self.mpDraw.draw_landmarks(img, handLms,
                self.mpHands.HAND_CONNECTIONS)
                return img
        def findPosition(self, img, handNo=0, draw=True):
                lmList = []
                if self.results.multi_hand_landmarks:
                myHand = self.results.multi_hand_landmarks[handNo]
                for id, lm in enumerate(myHand.landmark):
                # print(id, lm)
                h, w, c = img.shape
                cx, cy = int(lm.x * w), int(lm.y * h)
                # print(id, cx, cy)
                lmList.append([id, cx, cy])
                if draw:
                cv2.circle(img, (cx, cy), 15, (255, 0, 255), cv2.FILLED)
                return lmList
def main():
        pTime = 0
        cTime = 0
        cap = cv2.VideoCapture(0)
        detector = handDetector()
        while True:
                success, img = cap.read()
                img = detector.findHands(img)
                lmList = detector.findPosition(img)
                '''if len(lmList) != 0:
                print(lmList[4])
                cTime = time.time()
                fps = 1 / (cTime - pTime)
                pTime = cTime
                cv2.putText(img, str(int(fps)), (10, 70),
                cv2.FONT_HERSHEY_PLAIN, 3, (255, 0, 255), 3)'''
                cv2.imshow("Image", img)
                cv2.waitKey(1)
if __name__ == "__main__":
main()
import cv2
import time
import os
import hand_track as htm
import serial
import time
wCam, hCam = 640, 480
cap = cv2.VideoCapture(0)
cap.set(3, wCam)
cap.set(4, hCam)
pTime = 0
detector = htm.handDetector(detectionCon=0.75)
tipIds = [4, 8, 12, 16, 20]
#serialcomm = serial.Serial('COM7',9600)
#serialcomm.timeout = 1
while True:
        success, img = cap.read()
        img = detector.findHands(img)
        lmList = detector.findPosition(img, draw=False)
        # print(lmList)
        if len(lmList) != 0:
        fingers = []
        # Thumb
        '''if lmList[tipIds[0]][1] > lmList[tipIds[0] - 1][1]:
        fingers.append(1)
        else:
        fingers.append(0)'''
        # 4 Fingers
        for id in range(1, 5):
        if lmList[tipIds[id]][2] < lmList[tipIds[id] - 2][2]:
        fingers.append(1)
        else:
        fingers.append(0)
        # print(fingers)
        totalFingers = str(fingers.count(1))
        print(totalFingers)
        print(type(totalFingers))
        else :
        totalFingers = 0
        print(totalFingers)
        #h, w, c = overlayList[totalFingers - 1].shape
        #img[0:h, 0:w] = overlayList[totalFingers - 1]
        #cv2.rectangle(img, (20, 225), (170, 425), (0, 255,
        0),
        cv2.FILLED)
        #cv2.putText(img, str(totalFingers), (45, 375),
        cv2.
        FONT_HERSHEY_PLAIN,10, (255, 0,0), 25)'''
        cTime = time.time()
        fps = 1 / (cTime - pTime)
        pTime = cTime
        cv2.putText(img, f'FPS: {int(fps)}', (400, 70),
        cv2.
        FONT_HERSHEY_PLAIN, 3, (255, 0, 0), 3)
        cv2.imshow("Image", img)
        cv2.waitKey(1)
# serialcomm.write(totalFingers.encode())
# time.sleep(0.5)
#serialcomm.close()
```

Fig. 4 Codes for gesture recognition model

3 Results and Discussions

The designed model can observe and recognize the different hand movements and patterns per the specified set of rules. As discussed above, there is a precise key point localization of 21 3D hand coordinates, and a key point is recognized only when they are at their desired positions. If the key point associated with the top of the index finger gets below the key point related to the middle of the index finger, in case the finger is bent. In those cases, the gesture is not recognized. Figure 5a–e represents the outputs of the developed gesture recognition setup using an LED system.

It can be seen from Fig. 5a that there is no output if all fingers are bent. Corresponding to the index finger (1 finger), the first LED illuminates, as shown in Fig. 5b. Similarly, LED 2 illuminates corresponding to 2 fingers (index + middle), as shown in Fig. 5c. Inputs of 3 fingers (index + middle + ring) and 4 fingers (index + middle + ring + pinky) correspond to the illumination of LED 3 and LED 4, respectively, as shown in Fig. 5d and e.

Hand Gesture is a better alternative to express gesture recognition system-based signals into some related events. Therefore, these provide natural, innovative and

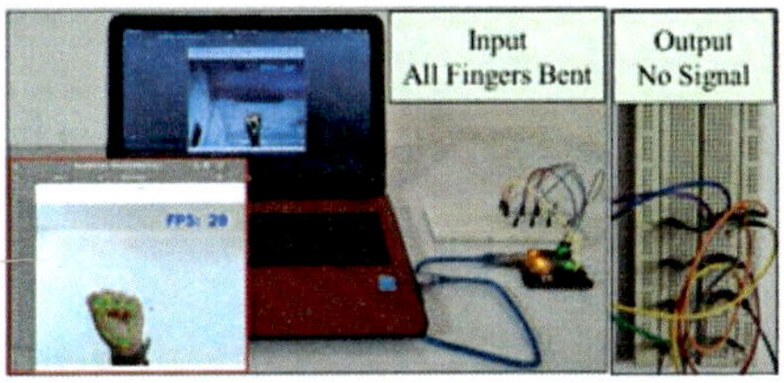

(a). Output for all fingers bent.

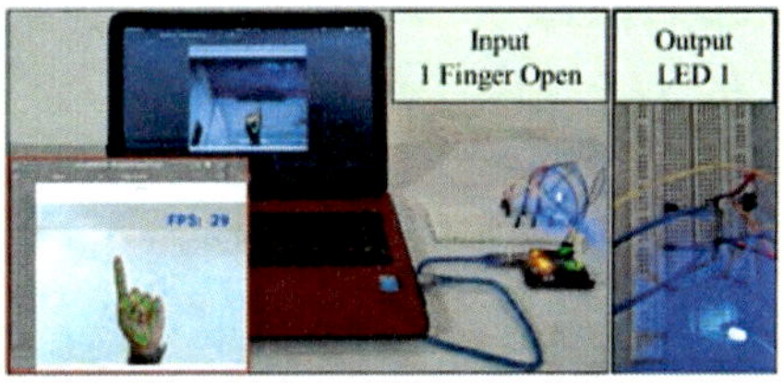

(b). Output for one open finger.

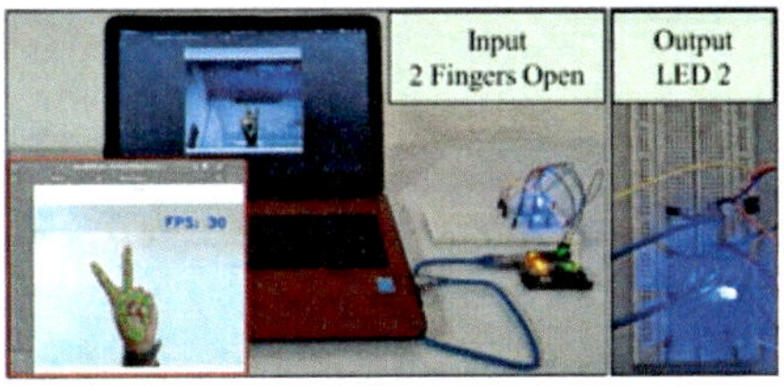

(c). Output for two open fingers.

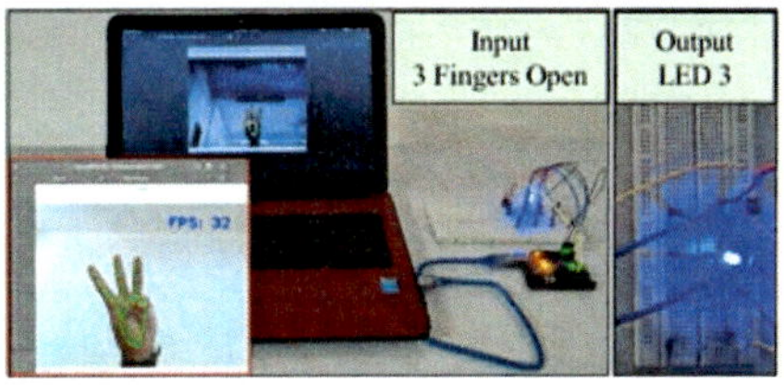

(d). Output for three open fingers.

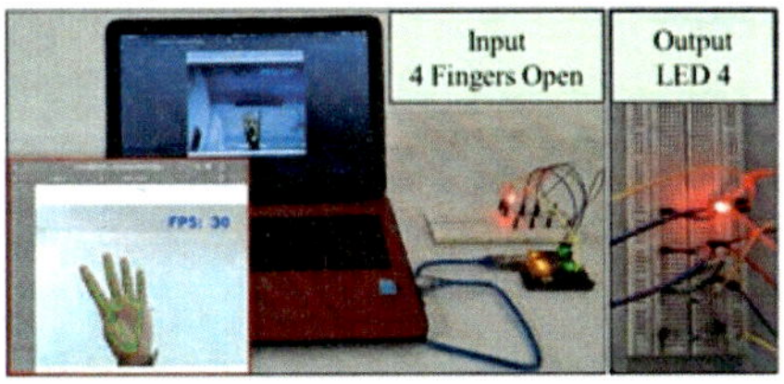

(e). Output for four open fingers.

Fig. 5 **a** Output for all fingers bent. **b** Output for one open finger. **c** Output for two open fingers. **d** Output for three open fingers. **e** Output for four open fingers modern non-verbal communication between humans and machines. These outputs can be used in gesture-based gaming control, home appliance automation, gesture-based car driving, and biomedical applications.

4 Conclusions

The following conclusions have been drawn based on the above study:

- A vision-based hand gesture recognition setup has been developed and modelled for hand gesture recognition.
- The developed setup has been tested for different gesture outputs and found that the developed setup is capable of recognizing different gestures appropriately.
- The developed setup has wide applications in gaming, automation and health industries.

References

1. Wachs JP, Kölsch M, Stern H, Edan Y (2011) Vision-based hand-gesture applications. Commun ACM 54:60–71
2. Mitra S, Acharya T (2007) Gesture recognition: a survey. Syst Man Cybern Part C: Appl Rev 37:311–324
3. Erol A, Bebis G, Nicolescu M, Boyle RD, Twombly X (2007) Vision-based hand pose estimation: a review. Comput Vis Image Underst 108:52–73
4. McNeill D (2000) Language and gesture, vol 2. Cambridge University Press
5. Damdoo R, Kalyani K, Sanghavi J (2020) Adaptive hand gesture recognition system using machine learning approach. Biosci Biotechnol Res Commun 13(14):106–110
6. Grobel K, Assan M (2002) Isolated sign language recognition using hidden Markov models. Comput Cybern Simul IEEE 1:162–167
7. Zhu Y, Xu G, Huang H (2000) Dynamic isolated gesture recognition based on appearance. J Softw 11(1):54–61
8. Ling X, Renfa L, Fanzi Z, Weilan Q (2013) Dynamic gesture recognition method based on learning sparse representation. J Commun 24(6):128–135
9. Wang K, Yu H, Zhang P (2012) Real-time gesture recognition based on AdaBoost algorithm and optical flow matching. Microelectron Comput 5(29):138–141
10. Kavana KM, Suma NR (2022) Recognization of hand gestures using MediaPipe hands. Int Res J Modernization Eng Technol Sci 4(6):4149–4156
11. Google: https://developers.google.com/mediapipe/solutions/vision/hand_landmarker

Stochastic Petri Nets Maintenance Modeling of a Butter Oil Processing Plant

Satnam Singh, Ankur Bahl, and Anish Sachdeva

Abstract Preventive Maintenance (PM) has been playing a key role in the reliability and sustainability of industrial operations by preventing the failure of various components resulting in a breakdown. Further, this breakdown in industrial operations inhibits the scale and goals of production thereby affecting financial growth. Butter oil is one of the most consumed oils in the world, so its production in processing plants should be monitored and optimized maintenance schedules for the equipment should be adopted which affects the production in a plant. In the present paper, the preventive maintenance (PM) modeling of the butter oil processing plant is carried out using Stochastic Petri nets (SPN). The optimal preventive maintenance schedules are decided under the various constraints of maximizing availability, minimizing cost, and maximizing revenue generation. The programming package Mathematica is used to solve the complex equations of the framework in maximizing the availability, revenue per unit, and minimizing cost. The efficacy of the model is assessed by performance parameters of plant availability and maintenance cost. It is found from investigations that the system availability could be increased from 0.81 to 0.91 and the cost could be reduced from 40 to 32%. Hence, it is concluded that the implementation of preventive maintenance schedules could be useful in achieving production goals.

Keywords Petri Nets · Availability · Optimal Schedule · Preventive Maintenance

S. Singh · A. Bahl (✉)
School of Mechanical Engineering, Lovely Professional University, Phagwara, Punjab 142111, India
e-mail: ankur.bahl@lpu.co.in

A. Sachdeva
Department of Industrial and Production Engineering, Dr. B R Ambedkar National Institute of Technology, Jalandhar 144027, India

A. Sachdeva et al. (eds.), *Recent Advances in Operations Management and Optimization*, Lecture Notes in Mechanical Engineering,
https://doi.org/10.1007/978-981-99-7445-0_23

1 Introduction

Time degradation of components of the system results in system failure. Preventive maintenance (PM) is used to slow down this degradation process which results in the extended life cycle of the system and enhances the system's reliability. Moreover, maintenance helps to keep the system assets in a fit state and perform their intended function. Therefore, maintenance has a major impact on quality and achieving production goals. The purpose of maintenance is to bring back the failed components to their original state. To do this, mainly two maintenance policies are being followed in the industry such as corrective maintenance (CM) policy and PM policy. Effective implementation of PM requires following optimal preventive maintenance schedules. Various research addresses the issue of finding optimal maintenance schedules [1–7]. Optimization of the preventive maintenance schedule was done by setting up one of the factors such as life cycle cost, reliability, availability, and cost as constraints to the model [1, 8–12]. Further, the genetic algorithm (GA) technique was used with the help of the MATLAB interface for calculating the PM schedules considering the availability as a constraint [13, 14]. Bahl et al. [15] discussed the application of Petri nets in the availability analysis of distillery plants. Penttinen et al. [16] carried out a risk assessment of large and complex systems with dynamic behavior. Kumar et al. [17] carried out the Petri net-based simulation modeling of a complex industrial paint manufacturing unit. Considering the importance of implementation of preventive maintenance in various industries, in this work, the preventive maintenance model of the butter oil processing plant based on the Petri net model is presented. In the present Petri net model, the optimal PM schedules were found using a combined optimization model of maximizing availability, minimizing maintenance cost, and maximizing revenue generation. The complex equations are solved using the Mathematica programming package.

2 Introduction to Petri Nets

An ample amount of information related to Petri nets is available in the literature [18]. Therefore, a brief overview of the basics of Petri nets is presented in this section. A Petri net is a directed graph with two disjoint sets of nodes places and transitions. Places represent the conditions for an event, whereas transition represents the event. They are represented graphically with a circle and rectangular box. The places and transitions relate to the help of arcs which are represented by arrowheads. An arc directed from places to transitions is called an input arc, whereas an arc that connects transition to place is known as an output arc. By default, an arc has been assigned the unit weight, however, the values of weight can change which represents a multiplicity of an arc. Each place is marked with a token which is denoted by a small dotted circle. A transition is enabled if all the input places carry the token and the transition fires resulting in the movement of the token from one place to another. Each transition

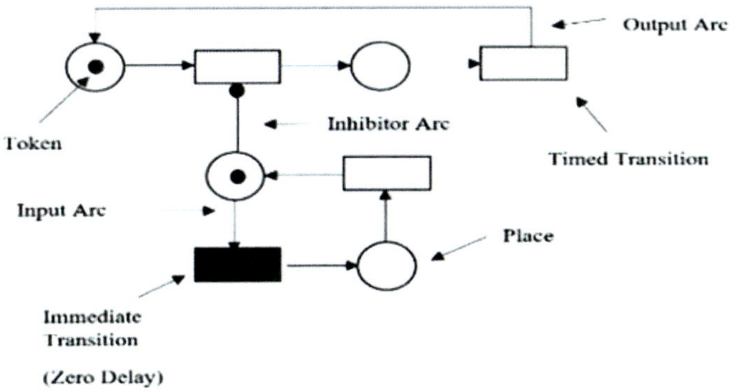

Fig. 1 Illustration of a simple Petri Net Mode

is associated a with time delay equal to failure and repair rates of the components. However, when the delay is zero, the transition is termed an immediate transition. A transition is disabled if one of its arcs is an inhibitor arc, therefore inhibitor arc restricts the movement of tokens from place to transition when the number of tokens in the place is equal to the multiplicity of the input arc. The illustration of a simple Petri net model is shown in Fig. 1.

3 Butter Oil System Description

Butter oil or ghee refers to the clarified butter fat obtained mainly from butter after removing the water and SNF (solids-not-fat) contents. It is the richest source of milk fat and is prepared either with butter or cream. The butter oil production system consists of four subsystems, i.e., heater, clarifier, filling, and granulation. Out of these, except for the granulation subsystem, all the units are subject to random failures. The flowchart of the butter oil-making process is shown in Fig. 2.

3.1 Sub-systems Description

3.1.1 Heater Unit

It consists of a kettle in which the temperature of butter is raised slowly with the help of steam. The final temperature is monitored to be not more than 107–110 °C until its color is reddish-brown. Ghee along with the residue can settle down for 25–30 min in the kettle before filtration. It consists of two units in series. Hence, if one unit fails, the system fails.

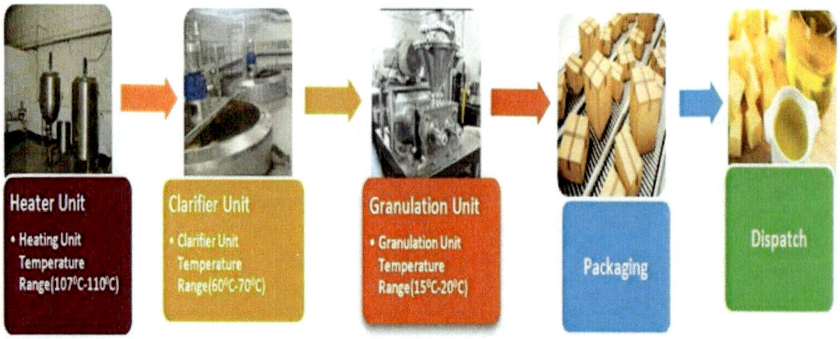

Fig. 2 Schematic of butter oil plant and the processes

3.1.2 Clarifier Unit

Clarification is carried out at around 70 °C to clarify all the residue particles from ghee. It consists of two units in parallel. Partial failure of this system reduces the capacity of the system. A major failure occurs only when both units fail.

3.1.3 Filling Unit

In this section, ghee tins are filled, weighed, and sealed simultaneously. The filling temperature is strictly watched to remain between 40 and 45 °C. There are two filling units. Failure of any one unit reduces the working capacity while the system completely fails when both units break down.

3.1.4 Granulation Unit

This subsystem consists of a refrigerating unit where the temperature is maintained between 15 and 20 °C. This section rarely fails and hence has not been considered for analysis.

The butter oil production system consists of four subsystems, i.e., heater, clarifier, filling, and granulation. Out of these, except for the granulation subsystem, all the units are subject to random failures. There are two units for each of the heating, clarifier, and filler. The two heating units are connected in series and they must run for a system to be operational. The two units the clarifier and filling units are in a parallel arrangement. Failure of any one of the units of these two leads to the running of the plant at reduced capacity. The repair of the unit restores the system to its full running capacity. The arrangements of units in the butter oil processing plant are shown in Fig. 3.

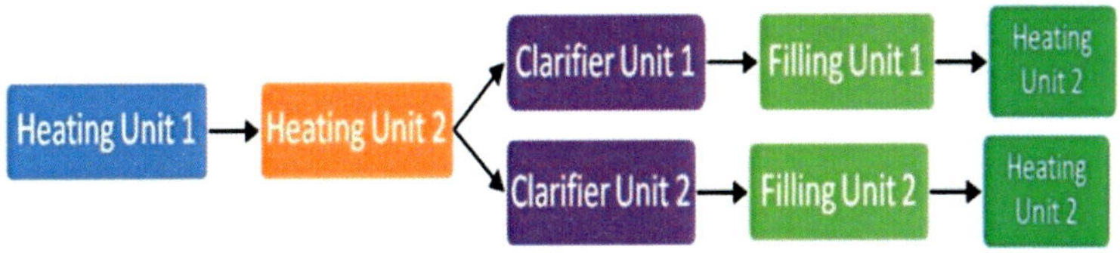

Fig. 3 Schematic arrangement of units of butter oil processing system

4 Preventive Maintenance Mathematical Model

Mathematica© software [19] is used to maximize plant availability, and revenue per unit and minimize the maintenance cost. However, the same can be done with the help of the following equations implemented in Mathematica.

The availability of the plant can be maximized using [20]

$$A_P = \text{Maximize}[\{A, \{\beta > (1 - (1 - \gamma) * e^{-\left(\frac{t}{\lambda}\right)^k})\&\&0 <= \beta <= 1\}\}, t] \quad (1)$$

Similarly, to minimize maintenance costs in a plant, the following equation is recommended [20]

$$\begin{aligned} C = \text{Minimize}[\{CP, \{\beta >; 1 - ((1 - \delta * \alpha) \\ + \mu * \alpha * (1 - \delta)) * e^{-\left(\frac{t}{\lambda}\right)^k} \&\&0 \le \beta \le 1\}\}, t] \end{aligned} \quad (2)$$

Further, revenue earning per unit can be maximized using

$$\begin{aligned} R = \text{Maximize}\left[\left\{P, \left\{\beta > \left(1 - \left(\frac{(1 - \alpha) + w * ((1 - \delta * \alpha) + \mu * \alpha * (1 - \delta))}{1 + w}\right)\right.\right.\right.\right. \\ \left.\left.\left.\left. * e^{-\left(\frac{t}{\lambda}\right)^k}\right)\&\&0 \le \beta \le 1\right\}\right\}, \{t\}\right] \end{aligned} \quad (3)$$

where

A_p: Represents the availability function under preventive maintenance.

C: Represents the cost function.

R: Represents the revenue earning per unit.

δ: Represents the maintenance cost ratio under PM and CM policies.

α: Represents the ratio of repair time under PM and CM policies.

β: Represents the ratio of MTBF under PM to the MTBF under CM.

μ: Represents the ratio of repair time to failure time under CM policy.

Table 1 Decision parameters for units of butter oil processing plant

Units	Decision parameters					
	Scale factor (θ)	Shape factor (β)	δ	α	μ	ω
Heating unit	72	2.1	0.4	0.50	0.12	0.1
Clarifier unit	77.1	2.72	0.45	0.41	0.07	0.08
Filling unit	76.4	2.76	0.5	0.37	0.04	0.06

Table 2 The optimal preventive maintenance schedule for various units

Units	T_A	T_C	T_P	T^*
Heating unit	73.9	49.23	72.4	51.45
Clarifier unit	58.36	34	54.76	39.48
Filling unit	52.5	35.7	49.78	51.65

w: Represents the ratio of maintenance cost per unit time to net earnings per unit time.

k : Represents the shape factor under Weibull distribution.

λ: Represents the scale factor under Weibull distribution.

Further, the Mathematica and Petri net models are not integrated into the same interface. However, the optimized preventive maintenance time calculated using the above-mentioned equation would be calculated using Mathematica given as input to the Petri net Models. Similarly, such a method can be implemented in different industrial applications where optimizing circumstances is expected.

The decision parameters derived from maintenance data for units are given in Table 1. The ratios α and μ are obtained from the maintenance data. The ratios δ and ω have been finalized based on the subjective opinion of the maintenance experts. Using these decision parameters, PM schedules have been obtained for various units of a butter oil processing plant for availability T_A, maintenance cost T_C and revenue earning T_P criteria, respectively. The overall value of PM schedules T^* is determined from the model and is shown in Table 2.

5 Petri Net Model of Butter Oil Processing Plant

The Petri net model for preventive maintenance of a butter oil processing plant is shown in Figs. 4 and 5.

The detailed description of places and transitions is compiled in Table 3.

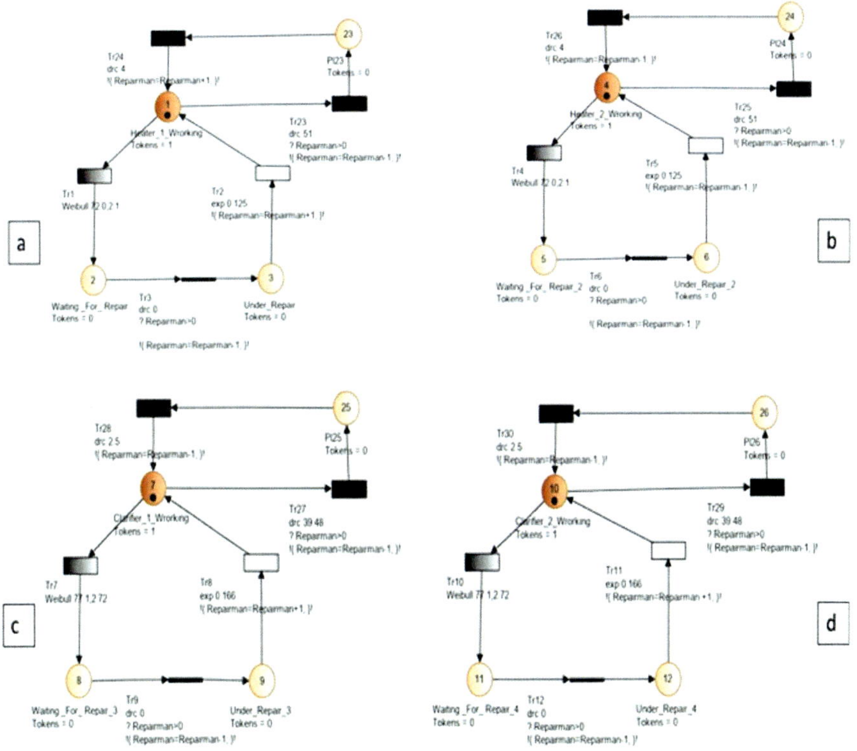

Fig. 4 Preventive maintenance-based Petri net model of butter oil processing plant. **a** Heating unit 1, **b** heating unit 2, **c** clarifier unit 1, **d** clarifier unit 2

6 Performance Analysis of the System

The performance analysis of the butter oil processing system is done using the MC simulation of the PN model using the Graphical Interface for Reliability Forecasting (GRIF 2023) Petri net module. The simulation evaluates the plant's availability. It runs for 10,000 replications for 1 year at a 95% confidence level. The system performance analysis is evaluated by considering the effect of preventive maintenance on plant availability. Also, the system is evaluated in terms of the plant working at reduced capacity. The data relating to the failure and repair times of all units of the system are shown in Table 4. The plant availability without and with preventive maintenance is shown Fig. 6.With the implementation of preventive maintenance, the system availability increased from 0.81 to 0.91.

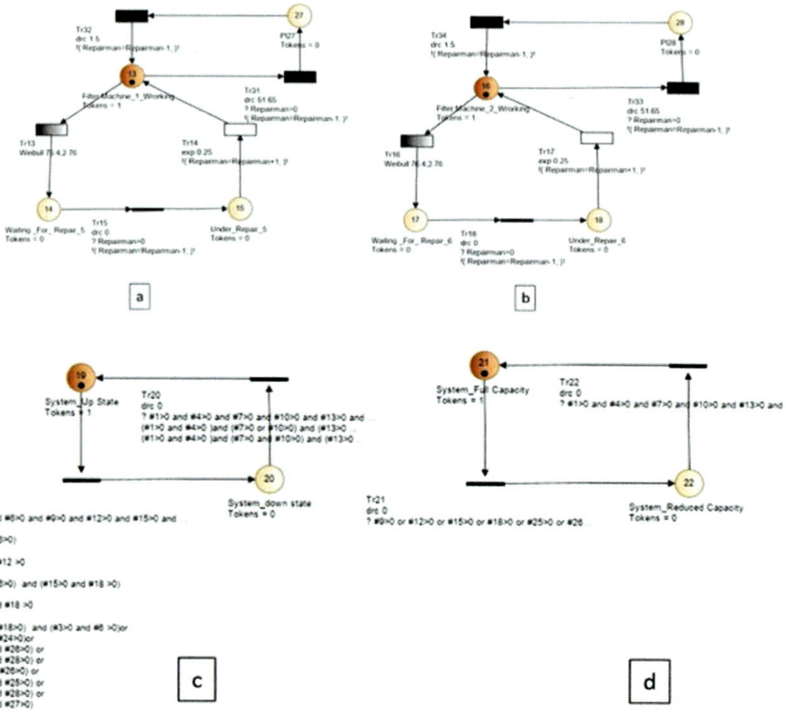

Fig. 5 Preventive maintenance-based Petri net model of butter oil processing plant. **a** Filler unit 1, **b** filler unit 2, **c** system down state, **d** system reduced capacity state

7 Cost Analysis

The maintenance cost is evaluated from the model and is shown in Fig. 7. It is clearly understood from the figure that the corrective maintenance cost is higher compared to the preventive maintenance cost in all the units such as the Heating unit, the Clarifier unit, and the Filler unit. However, there is a considerable reduction of around forty percent in the maintenance cost of the Filler Unit. The decrease in the maintenance cost is due to the implementation of optimal preventive maintenance schedules.

8 Conclusions

The present work is focused on the Petri net-based modeling of a butter oil processing plant. The GRIF Petri net module is used for the simulation to study the dynamic behaviour of the plant. The preventive maintenance policy is followed for all three units of the plant. It can be concluded from investigations that the system availability

Table 3 Description of places and transitions

Places	Description
Heater_1_Working and Heater_2_Working	Places represent the working states of heating units
Clarifier_1_Working and Clarifier_2_Working	Places represent the working states of Clarifier units
Filler_1_Working and Filler_2_Working	Places represent the working states of Filling units
Waiting_For_Repair_1 andWaiting_For_Repair_2	Places represent the heating unit has failed and is waiting for repair
Waiting_For_Repair_3 and Waiting_For_Repair_4	Places represent the clarifier unit has failed and is waiting for repair
Waiting_For_Repair_5 and Waiting_For_Repair_6	Places represent the Filler unit has failed and is waiting for repair
Under_Repair_1 and Under_Repair_2	Places represent the heating unit is under repair
Under_Repair_3 and Under_Repair_4	Places that represent the clarifier unit is repair
Under_Repair_5 and Under_Repair_6	Places represent the clarifier unit is under repair
System_Full_Capacity	Place represents that all the units are working
System_Reduced_Capacity	Place represents that some of the units are not working and the system is running with reduced capacity
System_Up_State	Place represents that the system is in a working state
System_Down_State	Place represents that the system is in the downstate or the system is not available
Transition	Description
Tr1, Tr4, Tr7, Tr10, Tr13, Tr16	Represents the timed transitions associated with stochastic delay in all the units and firing of these transitions leads to failure of the corresponding unit. The scholastic delays are related to the failure rates of the units
Tr2, Tr5, Tr8, Tr11, Tr14, Tr17	Represents the timed transitions associated with stochastic delay in all the units and firing of these transitions leads to repair of the corresponding unit. The scholastic delays are related to the repair rates of the units
Tr27, Tr28	Represents the timed transition having a period equal to the optimal preventive maintenance schedule for the heating unit
Tr29, Tr30	Represents the timed transition having a period equal to the optimal preventive maintenance schedule for the clarifier unit
Tr31, Tr32	Represents the timed transition having a period equal optimal preventive maintenance schedule for the Filling unit

Table 4 Failure and repair data for various units of the butter oil processing system

Units	Failure data (Weibull distribution)			Repair data (exponential distribution)
	Scale factor (θ)	Shape Factor (β)	MTBF (hrs)	MTTR (hrs)
Heating unit	72	2.1	65	8
Clarifier unit	77.1	2.72	76	6
Filling unit	76.4	2.72	82	4

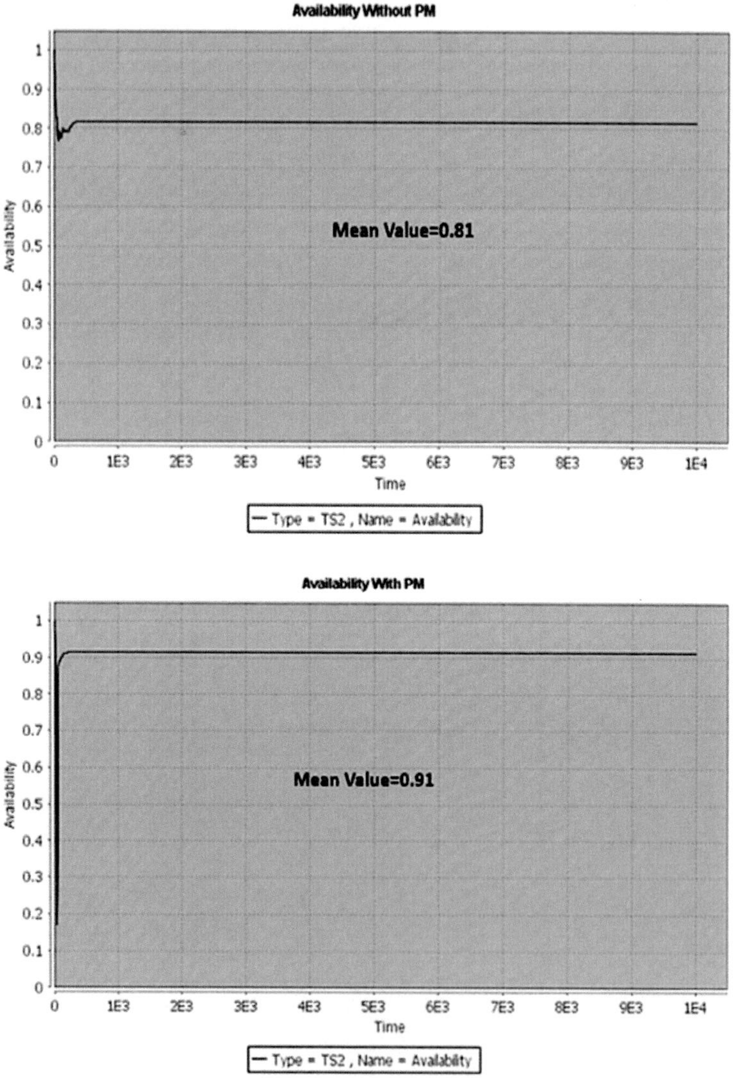

Fig. 6 Availability of plant without and with preventive maintenance

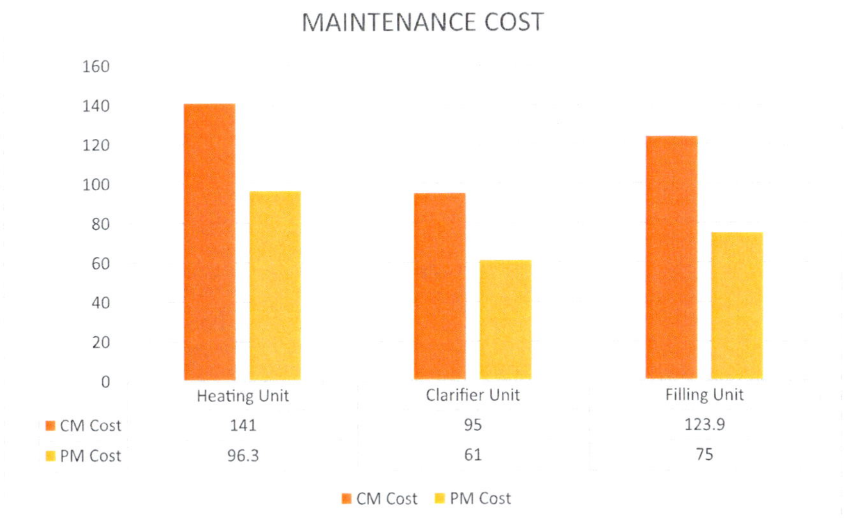

Fig. 7 Comparison of maintenance cost

increased from 0.81 to 0.91 (a 10% increase) and the cost was reduced from 32 to 40%. Hence, it is also concluded that the implementation of such preventive maintenance could be useful in achieving production goals and bridging the gap between supply and demand.

References

1. Chou JS, Le TS (2001) New methods to minimize the preventive maintenance cost of series—parallel systems using ant colony optimization. Reliab Eng Syst Saf 4:1402–1410
2. Dekker R, Scarf PA (1998) On the impact of optimisation models in maintenance decision making: the state of the art. Reliab Eng Syst Saf 60:111–119
3. Hipkin IB, Lockett AG (1995) A study of maintenance technology implementation. Omega
4. Jayabalan V, Chaudhuri D (1992) Cost optimization of maintenance scheduling for a system with assured reliability. IEEE Trans Reliab
5. Nagakawa T (1980) A summary of imperfect preventive maintenance policies with minimal repair. Res Oper Oper Res 14:249–255
6. Valdez-Flores C, Feldman RM (1989) A survey of preventive maintenance models for stochastically deteriorating single-unit systems. Nav Res Logist
7. Wang H (2002) A survey of maintenance policies of deteriorating systems. 139:469–489
8. Gopalakrishnan M, Mohan S, He Z (2001) A tabu search heuristic for preventive maintenance scheduling. Comput Ind Eng
9. Ntuen CA (1991) An economic preventive maintenance scheduling model with truncated gamma function. 31:31–38
10. Nakamura M, Member S, Katafuchi T, Hatazaki H (2001) Decisions for maintenance-intervals of equipment in thermal power stations. Based Few Data 50:360–364

11. Beichelt F, Beichelt F (2001) A replacement policy based on limiting the cumulative maintenance cost
12. Vassiliadis CG, Pistikopoulos EN (2001) Maintenance scheduling and process optimization under uncertainty. 25:217–236
13. Bris R, Châtelet E, Yalaoui F (2003) New method to minimize the preventive maintenance cost of series-parallel systems. Reliab Eng Syst Saf
14. Lapa CMF, Pereira CMNA, de Barros MP (2006) A model for preventive maintenance planning by genetic algorithms based in cost and reliability. Reliab Eng Syst Saf
15. Bahl A, Sachdeva A, Garg RK (2018) Availability analysis of distillery plant using petri nets. Int J Qual Reliab Manag 35:2373–2387
16. Penttinen JP, Niemi A, Gutleber J, Koskinen KT, Coatanéa E, Laitinen J (2019) An open modelling approach for availability and reliability of systems. Reliab Eng Syst Saf 183:387–399
17. Kumar A, Kumar V, Modgil V, Kumar A (2022) Stochastic Petri Nets modelling for performance assessment of a manufacturing unit. Mater Today Proc 56:215–219
18. Murata T (1989) Petri nets: properties, analysis and applications
19. Wolfram Research (2019) Mathematica
20. Sachdeva A, Kumar D, Kumar P (2008) Planning and optimizing the maintenance of paper production systems in a paper plant. Comput Ind Eng 55:817–829

Developing Artificial Neural Network Based Model for Backorder Prediction in Supply Chain Management

Aarti Rana and Rajiv Kumar Sharma

Abstract In today's golden era of digitization, every industry has been focused on the adaptation of new technologies to boost their market dominance and company revenue. For every industry, supply chain management is the main area of concentration in order to develop brand value and seize local and global markets. Supply chain management (SCM) is progressively becoming a pillar for any firm. It is the top-down and bottom-up centralized control of the flow of goods, information, services, and money. It encompasses all processes, from acquiring raw materials to manufacturing, marketing, regulating supply and demand, and creating final products in response to market needs. Industries may be able to reduce excess costs associated with their products and deliver the products to customers in a more effective and efficient manner by managing proper supply chain operations. The primary customer demand is for a high-quality product at a reasonable price and with prompt delivery. To satisfy customer demand and enhance profit by capturing market share, companies have to focus on digitization in supply chain management. Digitization aids in forecasting demand fluctuations, reducing unnecessary inventory management and managing back orders. Backorders, improper data handling, unexpected delays in delivering products, not being able to forecast demand properly, and the bullwhip effect are all major threats to the overall supply chain profitability. In this study, the main focus is the usage of artificial neural networks to predict backorder in supply chain management to overcome demand fluctuations.

Nomenclature

SVR/SVM	Support Vector Regression/Support Vector Machine
GBML	Gradient Boost Machine learning
ARIMA	Autoregressive Integrated Moving Average

A. Rana (✉) · R. K. Sharma
Department of Mechanical Engineering, National Institute of Technology, Hamirpur 177005, India
e-mail: 21mme201@nith.ac.in

A. Sachdeva et al. (eds.), *Recent Advances in Operations Management and Optimization*, Lecture Notes in Mechanical Engineering,
https://doi.org/10.1007/978-981-99-7445-0_24

CART	Classification and Regression Tree
MAE	Mean Absolute Error
AUC	Area Under the Curve
KNN	K Nearest Neighbour
BO	Backorder

1 Introduction

Backorder typically means that a company's ability to meet client demand for a specific item or service is inadequate. A solid sales performance could cause a product backorder because of the high demand for that particular product, but companies will not be able to achieve that market demand completely. Poor safety stock level creates a stock situation for that particular product, known as a backorder. If that "stock out" kind of situation happens continuously it may cause for loss in customer satisfaction. Instead of that, if the product which is not in demand and you are keeping the inventory, it will unnecessarily increase the SC cost and pull down the profitability of the SC. So, a proper review of safety stock is necessary. Market volatility is a major cause of backorder. Volatility has always played a crucial role in supply chains. In supply chain management, demand for a product such as woolen cloths, and coffee. increases drastically, because winter is approaching—or perhaps at the same instant time the demand for cold drinks, etc. decreased significantly. In change believes it is important that companies make conscious choices about how to deal with volatility and make that a part of how they manage their business [1]. Unplanned changes to upstream and downstream material flows that lead to a mismatch between supply and demand at the target firm are referred to as "supply chain instability". Even the most exact calculations won't promise that everything will go as planned, hence regulating supply chain volatility is frequently cited as one of the biggest problems of modern supply chain management. Customers will be dissatisfied, and competitors will lose money if they don't have enough inventory to meet the increased demand.

Companies must use strong supply chain strategies to effectively manage volatility in a demand-driven economy, which is a big concern. The focus is usually on an all-encompassing systemic approach, which makes SC more efficient. The focus should be on reduction in Lead time, visibility, transparency in supply chain and collaboration, lowering of complexity, buffer inventory management, shorten cycle time, etc.

In this study, an artificial neural network (ANN) is used for identifying the back-order prediction and for analyzing the results [2]. Deep learning tools are used for heavy datasets and to get more accurate results. Utilizing increases in computing power and improved training techniques, it integrated massive neural network data into several layers of processing units to discover complicated patterns in massive amounts of data. Here, this study does some analysis of those data to gain insights or data analytics. Data processing is the initial step for deep learning models; after that, move ahead with model training, testing, and analyzing.

2 Literature Review

The conventional approach is not a workable option to prevent backorder (BO) difficulties. artificial intelligence (AI) is employed, in which various appropriate algorithms are applied in accordance with the data structure, to optimize the time and obtain accurate results. Many researchers did work in the domain of backorder prediction in supply chain management by using machine learning algorithms, and only a few authors tried to use deep learning in the out-of-stock product forecasting domain. There are some machine learning approaches such as SVR, NN, SVM, GBM, RBF, CART, and Ensemble-DRF. and deep learning that can be utilized in inventory management [3]. Andrés Martnez et al. have developed powerful analytics tools that predict consumer opinions by applying the Gradient Tree Boosting, Lasso and Extreme Learning Classifier machine learning techniques in non-contractual situations. Gradient boosting yields better outcomes [4]. Boosting is a popular ensemble classifier in which many slow classifiers are combined to create a stronger learner [5].

Xiang Wan et al.'s investigate the inverse correlation between inventory turnover and product variety [6] If product variety is greater than the amount required to manage adequate inventory and fulfil customer demand on time.

Studies by Maryam M. Najafabadi1 and others on the use of deep learning and problems with big data analytics [7]. Deep learning Utilizing increases in computing power and improved training techniques, it integrated massive neural network data into several layers of processing units to discover complicated patterns in massive amounts of data [8].

Scott Lee et al. [9] attempted to classify interpretable few-shot images using neural-backed decision trees. It's a powerful nonlinear regression and classification method influenced by theories about how the brain functions.

The predictability of material backorders in inventory management has been examined by Rodrigo Barbosa de Santis et al. by using machine learning [10]. A few products were recognized by the inventory model for the supply chain as having a greater incidence of shortages compared to demand and availability, leading to backorder problems. They researched different ML classifiers like CART and LOGIST in order to build a prognosis framework to solve the imbalanced rank problem.

Pak and N. Nahavandi B. Bagheri used an improved meta-heuristic algorithm to build a multi-objective green supply chain network for an automobile company [11].

To predict insurance premiums, Yi Yang et al. developed a gradient tree-boosted Tweedie compound Poisson model [12]. Samiul Islam and Saman Hassanzadeh Amin focused on the use of gradient boosting and distributed random forests machine learning algorithms for foreseeing supply chain backorder issues [13]. Future development will be based on the author's preferences, which include cost-cutting techniques, negative values in SC, flexible inventory control and BO scenarios in IM.

A blood supply chain management case study is described in which Babak Abbasi et al. make an effort to predict responses to complex optimization problems using machine learning [14]. The study looked at the positive effects of using ML approaches as tools for management in the network of blood supply facilities. They selected and applied well-known machine learning (ML) approaches like KNN, RF, CART, and ANN to address operational issues and come up with the best conclusions. The findings from the tests were successfully applied to find the best responses, leading to well-structured choices all over the network of hospitals.

The research conducted on performance versus interpretability trade-off by George Baryannis et al. was focused on using machine learning to anticipate supply chain issues [15]. Machine learning connects (Supply Chain Risk Management) SCRM aims to prioritize forecasting over intelligibility, or vice versa, with the machine learning algorithms. The real-world's case study of an aerospace manufacturing supply chain with increased delivery risk served as the inspiration for this work.

"Manufacturers' shipments, inventory," from Canada Statistics, according to Carbonneau et al., includes data on monthly sales [16]. Based on the research, this Canadian industry contains enterprises whose core business is employing NN and SVM algorithms to pour molten steel into investment moulds or full moulds to produce steel castings.

Wang Guanghui investigated SC demand forecasting using the SVR method and concluded that SVR outperforms SVM in terms of prediction performance based on accuracy and relative mean square error [17]. During the manufacturing process, they placed a strong emphasis on individualization and strict product customization.

3 Methodology

To predict backorders in supply chain management, one can use an Artificial Neural Network (ANN) process (Fig. 1). By implementing this process, supply chain managers can use ANN to make informed decisions about inventory levels and replenishment strategies, ultimately reducing the risk of backorders. In this analysis process, ANN classifiers are used in 2 different ways. In the first process, there is one input unit with 11 hidden units and output unit as shown in Fig. 4. In this process, the ANN model is not able to handle data properly. As data is highly imbalanced, and values of positive for backorder prediction is very less, So ANN model is not able to collect those positive values and the fitting is only done via negative values as shown in confusion metrics in Fig. 5a. Random resampling, the simplest method, entails selecting examples for the randomly transformed dataset. Oversampling and undersampling are the two primary approaches to random resampling for unbalanced

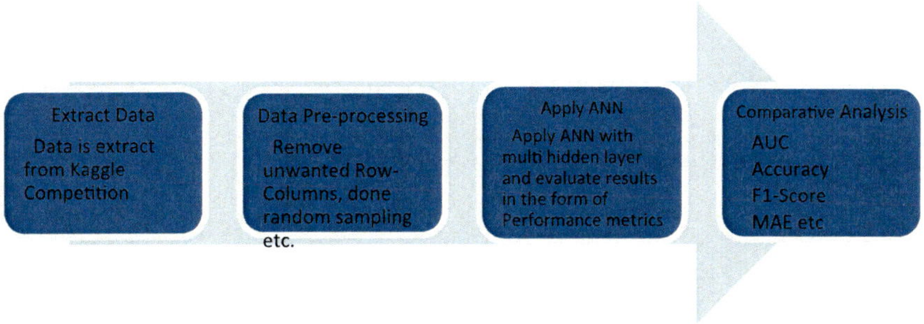

Fig. 1 Flow- Process for using ANN Classifier during Backorder prediction

classification. Random oversampling involves arbitrarily selecting occurrences from the minority class, then replacing them, and introducing them to the training dataset. Taking examples from the majority class at random and removing them from the training collection is known as undersampling.

Undersampling, or removing examples from the majority class, and oversampling, or duplicating examples from the minority class, are the two primary strategies for randomly resampling an unbalanced dataset.

So, to overcome those problems, the second model is used. In the second model, dropout with random undersampling is used to overcome those kinds of overfitting problems, as illustrated in the diagram. The steps involved in this process are:

- Data collection: Gather relevant data such as past sales, inventory levels, supplier lead times, demand forecasts, etc. [18]. In this study, it is from Kaggle competition.
- Data preparation: To eliminate any erroneous or unnecessary information, clean up and pre-process the data.
- Feature selection: Select the most important variables that have a significant impact on backorders. In this study, to predict backorder (output variable) the input variables are- national inventory, lead time, in transit product quantity, 9-months sales, 9-month forecast, mini bank (minimum stock required), and local backorder quantity.
- Model training: Train the ANN model using the selected features and historical data. This is done using a supervised learning algorithm such as backpropagation. After training the model, the fitting is done by using various batch sizes and epochs. The best batch size and epochs can be calculated by different codes and then the batch size with the epoch that provides the best accuracy is used.
- Model evaluation: Evaluate the effectiveness of the model using metrics such as recall, accuracy, precision, AUC, mean absolute error and F1 score[19]. The comparative values are seen in Fig. 3 and Table 2 for both analysis.
- Model deployment/oversight and development: Utilize the model that was trained in an everyday context to forecast upcoming backorders and avoid shortages of merchandise. After that, keep an eye on the model's performance and adjust it as needed to improve accuracy (Fig. 2).

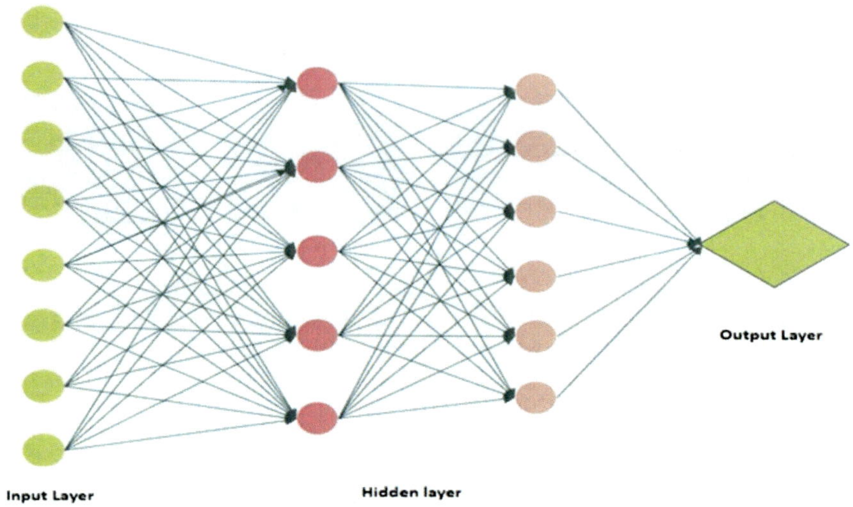

Fig. 2 Artificial neural network process structure

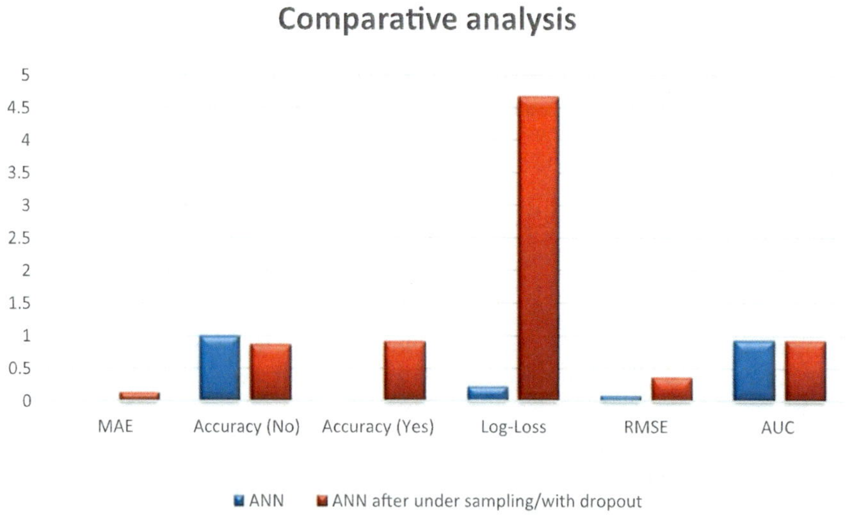

Fig. 3 Comparative analysis by various performance measures

4 Results and Analysis

As shown in Table 1 and Fig. 3, all values of the ANN model produce more precise results than a random sampling model, but the accuracy for yes outcomes is 0, indicating that the ANN model is incapable of detecting yes values. So, to overcome those issues, random sampling is done. Figure 5a also shows that the ANN model

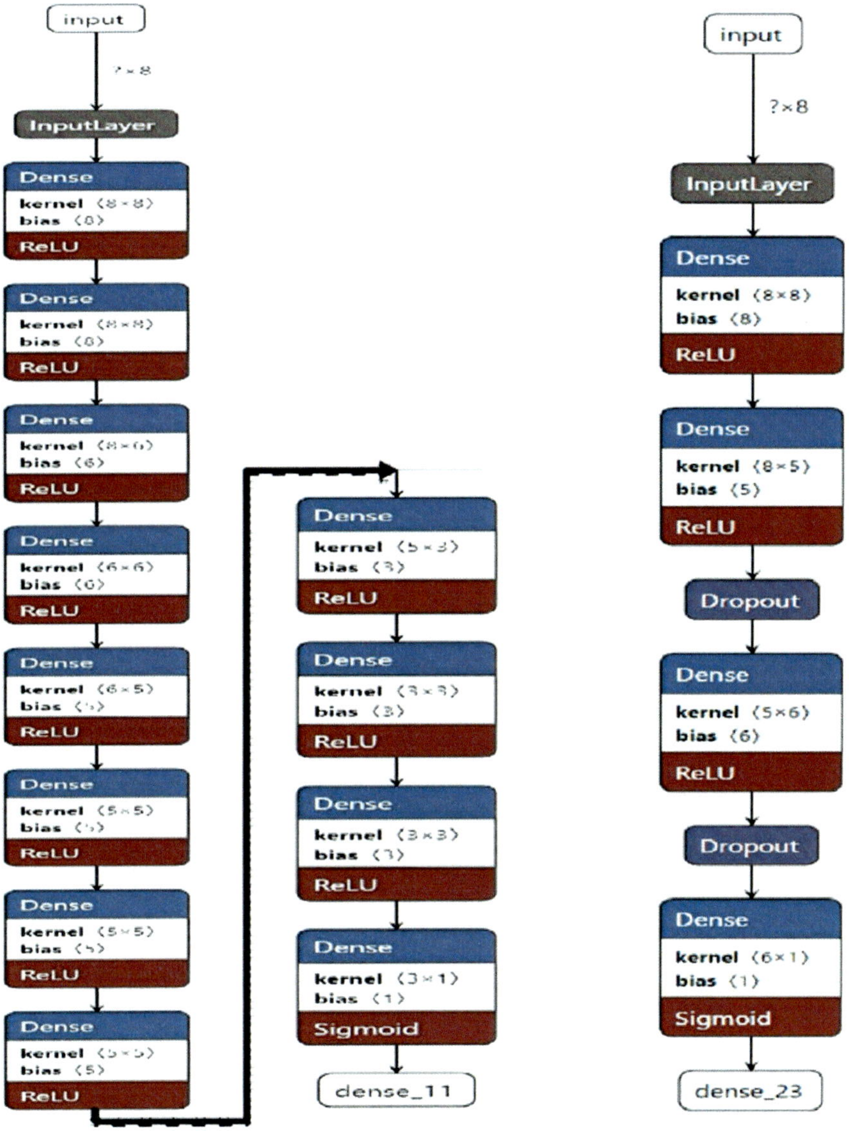

Fig. 4 Architecture for the proposed algorithm

is unable to predict the 1 (yes) value; values in the yes form are interpreted as 0 (wrong/no). The second model and the results shown in Fig. 5b solve this problem. Figure 6a and b show nearly identical AUC values for both models. Figure 7a and b show that after epoch number 2, accuracy has improved and losses have decreased. In this study, the batch size is 20 and the epochs are 5, and the validation split is 0.25 in the second analysis (random sampling), the batch size is 32, the epochs are 20,

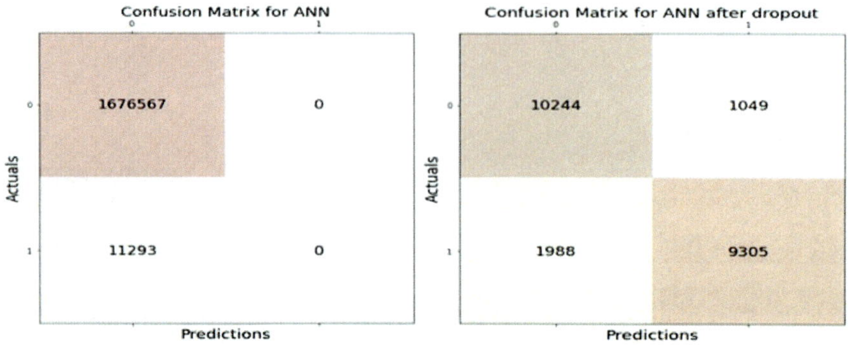

Fig. 5 **a** Confusion metrics without random sampling. **b** Confusion metrics with random sampling

and the call-backs are early stopping. Although random sampling is not as precise as ANN, it can balance our results.

Table 1 Validation values

Performance metrics	ANN	ANN after under sampling/with dropout
Mean absolute error	0.0066	0.13521
Accuracy for no	0.9933	0.86478
Accuracy for yes	0	0.91
Log-loss	0.2310	4.6702
Root mean square error	0.08179	0.36771
AUC	0.9323	0.92869

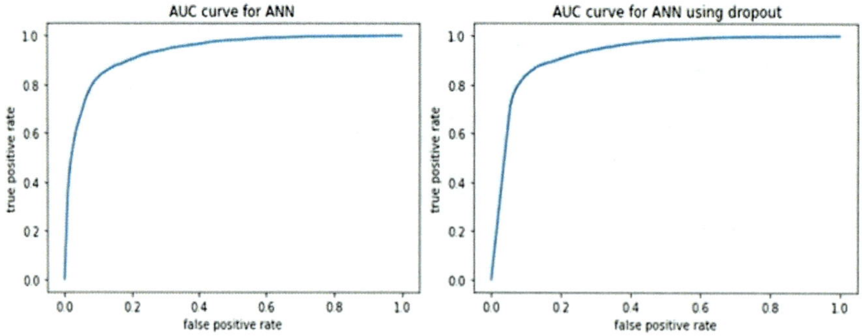

Fig. 6 **a** AUC curve without random sampling. **b** AUC curve with random sampling

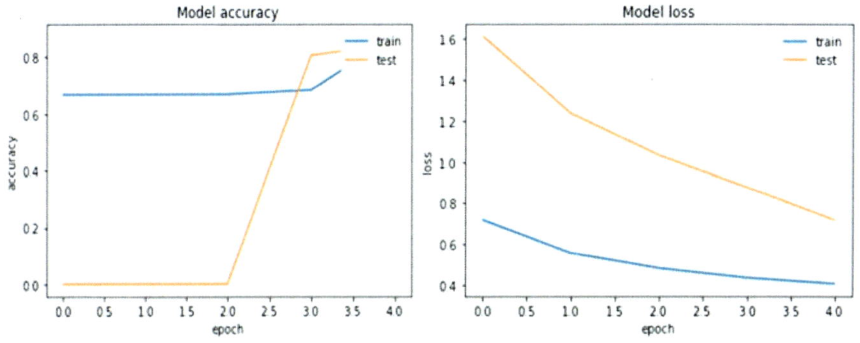

Fig. 7 **a** Accuracy with various epoch values. **b** Loss with various epoch values

5 Conclusions

The literature review shows that several studies have found that backorders are a real possibility in the field of supply chain management. ANN is a deep learning model, and it provides good results for heavy datasets. However, for a highly imbalanced dataset, balancing that dataset is necessary before applying ANN models. The results can be varied by increasing the hidden layers and altering the epoch and batch values. The study uses inventory, forecasted demand, sales, lead time, and other variables that have a significant impact on the backorder scenario to generate a realistic decision-making environment. In addition to this, backorder analysis uses a plethora of other factors. Backorder forecasting assists businesses in staying ahead of potential problems by alerting them to potential bottlenecks in their supply chains, allowing them to make necessary adjustments on time, all while preserving their reputations with consumers and shareholders through sustainable profit margins [20]. What will happen to the associated businesses if BO occurs? How much it can be avoided using AI methods, as well as the aspects of backorders that have a greater or lesser impact on backorder concerns, must all be thoroughly examined for that specific problem scenario.

References

1. Charis N, Kokkotis C, Serafeim, Elpiniki P (2021) An explainable machine learning pipeline for backorder prediction in inventory management systems; An explainable machine learning pipeline for backorder prediction in inventory management systems. https://doi.org/10.1145/3503823.3503866
2. Supply chain management and backorder products prediction utilizing neural network and naive bayes machine learning techniques in big data area: a real- life case study. https://www.researchgate.net/publication/363369600_Supply_Chain_Management_and_Backorder_Products_Prediction_Utilizing_Neural_Network_and_Naive_Bayes_Machine_Learning_Techniques_in_Big_Data_Area_A_Real-_life_Case_Study. Accessed 04 Nov 2022

3. Sarker IH (2021) Machine learning: algorithms, real-world applications and research directions. 2:160. https://doi.org/10.1007/s42979-021-00592-x
4. Martínez A, Schmuck C, Pereverzyev S, Pirker C, Haltmeier M (2020) A machine learning framework for customer purchase prediction in the non-contractual setting. Eur J Oper Res 281(3):588–596. https://doi.org/10.1016/j.ejor.2018.04.034
5. Biau G, Cadre B, Rouvière L. Accelerated gradient boosting
6. Wan X, Britto R, Zhou Z (2020) In search of the negative relationship between product variety and inventory turnover. Int J Prod Econ 222.https://doi.org/10.1016/J.IJPE.2019.09.024
7. Najafabadi MM, Villanustre F, Khoshgoftaar TM, Seliya N, Wald R, Muharemagic E (2015) Deep learning applications and challenges in big data analytics. J Big Data 2:1. https://doi.org/10.1186/s40537-014-0007-7
8. Seyedan M, Mafakheri F (2020) Predictive big data analytics for supply chain demand forecasting: methods, applications, and research opportunities. https://doi.org/10.1186/s40537-020-00329-2
9. Lee Joseph Gonzalez S, Matthew Wright E (2020) Interpretable few-shot image classification with neural-backed decision trees. Accessed 20 Oct 2022. [Online]. Available: http://www2.eecs.berkeley.edu/Pubs/TechRpts/2020/EECS-2020-71.html
10. de Santis RB, de Aguiar EP, Goliatt L (2018) Predicting material backorders in inventory management using machine learning. In: 2017 IEEE Latin American conference on computational intelligence, LA-CCI 2017—proceedings, vol 2017 November, pp 1–6. https://doi.org/10.1109/LA-CCI.2017.8285684
11. Pak N, Nahavandi N, Bagheri B (2022) Designing a multi-objective green supply chain network for an automotive company using an improved meta-heuristic algorithm. Int J Environ Sci Technol 19(5):3773–3796. https://doi.org/10.1007/S13762-021-03521-W
12. Yang Y, Qian W, Zou H (2016) Insurance premium prediction via gradient tree-boosted Tweedie compound Poisson models
13. Islam S, Amin SH (2020) Prediction of probable backorder scenarios in the supply chain using distributed random forest and gradient boosting machine learning techniques. J Big Data 7(1). https://doi.org/10.1186/s40537-020-00345-2
14. Abbasi B, Babaei T, Hosseinifard Z, Smith-Miles K, Dehghani M (2020) Predicting solutions of large-scale optimization problems via machine learning: a case study in blood supply chain management. Comput Oper Res 119.https://doi.org/10.1016/J.COR.2020.104941
15. Baryannis G, Dani S, Antoniou G (2019) Predicting supply chain risks using machine learning: the trade-off between performance and interpretability. Futur Gener Comput Syst 101:993–1004. https://doi.org/10.1016/J.FUTURE.2019.07.059
16. Carbonneau R, Laframboise K, Vahidov R (2008) Application of machine learning techniques for supply chain demand forecasting. Eur J Oper Res 184(3):1140–1154. https://doi.org/10.1016/J.EJOR.2006.12.004
17. Wang G (2012) Demand forecasting of supply chain based on support vector regression method. Procedia Eng 29:280–284. https://doi.org/10.1016/J.PROENG.2011.12.707
18. Predict products back-order to manage service level - dataset by amitkishore | data.world. https://data.world/amitkishore/can-you-predict-products-back-order. Accessed 02 Nov 2022
19. Orozco-Arias S, Piña JS, Tabares-Soto R, Castillo-Ossa LF, Guyot R, Isaza G (2020) Measuring performance metrics of machine learning algorithms for detecting and classifying transposable elements. Processes 8(6).https://doi.org/10.3390/PR8060638
20. Li Y (2017) Backorder prediction using machine learning for Danish craft beer breweries

Simulating Alternative Routes: A Model-Based Approach to Solve Traffic Congestion in Urban Areas

Vijay Itnal, Hritikesh Nilawar, and Ramkrishna Bharsakade

Abstract Indeed, the growing population and the escalating number of vehicles on the roads have given rise to a pressing concern in many large cities around the globe: traffic congestion. As urban areas become more densely populated and transportation options expand, the issue of traffic jams has become increasingly prevalent, impacting the daily lives of millions of people. Finding effective solutions to alleviate this problem has become a top priority for city planners and policymakers alike. Over the years researchers have suggested several multi-disciplinary methods such as Intelligent Transportation Systems (ITS), congestion pricing, improvements in public transit, promotion of alternative routes, prediction algorithms, and simulation techniques in literature. Many of these methods are based on real-time data and become more effective with technological advancements. This study is field field-based pilot study carried out in a major metropolitan city in India. The purpose of this study is to predict traffic congestion accurately and implement ground-level solutions with the help of simulation modeling in different scenarios. In this study, a simulation model was developed using a particular dataset, and the proposed solution model was tested and improved. Simulation of alternative routes has proven to be effective in reducing travel times, improving safety, and reducing environmental impacts. Promoting alternative routes has also shown potential to reduce congestion at specific points and improve public health by reducing emissions. Overall, a multi-disciplinary approach that considers various factors and ongoing evaluation is necessary for effective traffic congestion problem-solving.

1 Introduction

Traffic congestion is a growing problem in urban areas, causing longer travel times, air pollution and commuter stress. With an increase in population, it is becoming increasingly important to find effective solutions to this problem. Traffic congestion is

V. Itnal · H. Nilawar (✉) · R. Bharsakade
Vishwakarma Institute of Technology, Pune 411037, India
e-mail: hritikesh.nilawar19@vit.edu

A. Sachdeva et al. (eds.), *Recent Advances in Operations Management and Optimization*, Lecture Notes in Mechanical Engineering,
https://doi.org/10.1007/978-981-99-7445-0_25

a complicated problem that requires a multi-disciplinary approach to solve. Solutions can involve changes to transportation infrastructure, such as building new roads or expanding public transit systems, as well as changes to urban planning and land use.

There is a significant body of research focused on finding effective solutions to traffic congestion, including studies on the use of intelligent transportation systems, the implementation of congestion pricing, and the development of alternative modes of transportation such as biking and walking. This research aims to explore various approaches and strategies that can be employed to reduce traffic congestion and improve the overall transportation system efficiency in urban areas. The goal is to identify effective solutions that can be implemented in real-world settings and to provide insights for policymakers, urban planners, and transportation professionals in addressing this critical challenge.

The study provided focuses on resolving the traffic issues at a particular locality in a metro city of India. Hence, the traffic control system was proposed based on the simulations made with the use of PTV-Vissim software, this study is not only limited to streamlining the traffic but also to make benefits with respect to environmental aspect [1, 2]. Research is based on a comparative study of the current state model and future state model as proposed in the simulation.

1.1 Literature Review

Traffic congestion has been a domain of interest ever since the introduction of vehicles, many researchers published their works working to solve the traffic problems. It is important to understand every single attribute causing congestion. One of them is the capacity, how much the lane can occupy maximum to flow uninterruptedly, partial information of capacity to drivers may lead to worse conditions hence, it is essential to provide information from a valid source [3]. The capacity value is roughly estimated by N/T, where N is the total number of vehicles in the system, and T is the time taken by a vehicle to traverse a distance from one point to another [4].

With the aim to minimize the traffic on roads, it is important to spread awareness and support the public transport as in urban cities subsidies have been given to mass transit and justify it is economical [5]. Over the years, pollution and emissions were secondary reasons in order to resolve congestion in regions, even small changes in the speed of traffic will lead to major impacts of pollution on the environment this attribute was addressed by congestion mitigation strategies, speed management techniques, and traffic flow smoothing techniques [6]. Toll plaza was also studied as one of the major regions to investigate the air quality in nearby places, which was observed to be poor and impacts the health of infants and mothers nearby, E-ZPass was introduced in order to achieve a continuous flow of vehicles from the toll and the difference in air quality was recorded to have been improved [7].

Recent studies show that forecasting traffic congestion may lead to overcoming the issues beforehand, ARIMA model is one such technology that predicts congestion based on time-series analysis [8]. Another model was developed by improved kNN, where four kNN-based mechanisms have been introduced to detect traffic congestion [9].

Current developments in data science and machine learning and new technology advancements have led to solving this problem using improved prediction systems. Real-time data of traffic congestion performance extracted from AutoNavi map, which investigated spatial and temporal characteristics of urban traffic congestion in China [10]. Nowadays, simulation modelling is considered the best fit to solve congestion problems as it requires less time and capital before implementing in the real world. TCD algorithm and EB-TCD algorithm based on vehicle's speed or occupancy time-series data analysis, used to detect changes in traffic patterns while mitigating the effect of noise in the traffic data [11]. Another study uses four weeks of GPS data to create clusters of traffic congestion by Floating Car Data, as this framework allows more detailed analysis of different congestion clustering concerning the patterns of traffic, time frames of congestion and quantification of the strength of various congestion clusters [12].

2 Experimentation

2.1 Field Study/Methodology

The methodology for solving traffic congestion problems typically involves a multi-step process, which includes:

i. Data collection and analysis: The data was extracted from the observations for 1 month, on traffic patterns, road networks and other relevant environmental factors that lead to congestion and pollution. The data was sourced through video recordings and analyzed correspondingly.
ii. Modelling and simulation: The data collected earlier was organized and used for simulating the current scenario to predict different interventions which may impact the traffic flow. Simulation allows the researcher to test multiple scenarios simultaneously to evaluate their effectiveness before implementing the solution in the real world.
iii. Intervention development and testing: After developing an effective solution based on the results from the established model intervention can be made to reduce traffic congestion. Interventions may include the addition of new roads, flyovers, under bridges, improving road utilization and public transport.
iv. Evaluation and refinement: Once interventions have been implemented, their effectiveness can be evaluated through further data collection and analysis. This allows researchers to refine the interventions and make adjustments as necessary to achieve the desired results.
v. Implementation and monitoring: Finally, successful interventions can be implemented on a larger scale and monitored to ensure they continue to be effective over time. This may involve ongoing data collection and analysis to identify areas

for improvement or new interventions to address emerging traffic congestion issues.

Corresponding work was analyzed and the results were calculated, based on this the output was that only main roads were in use the majority of the time to pass through the region, as depicted in Figure 1. The traffic was densely concentrated and hence led to traffic congestion in the locality.

Interventions to the problem were made basically to avoid congestion and to decrease pollution levels at the junction. The solution gives a macro level output by engaging the side roads into daily routine, which were neglected by the commuters. Figure 2 shows how the concentration of vehicles at a node is distributed based on the proposed model of the scenario. In the proposed model, the traffic signal is also important to manage traffic and encourage citizens to use side roads to save time.

Overall, the methodology for solving traffic congestion problems involves an iterative process of data collection, modeling, intervention development, and evaluation, with the goal of identifying and implementing effective solutions to reduce traffic

Fig. 1 Current state road utilization

Fig. 2 Proposed utilization of roads

congestion and improve the overall transportation system efficiency in urban areas. Here, in this study, the focus is mainly on providing a feasible solution based on data collection and simulation, and a further approach will be carried out in the near future. Emissions from vehicles involve carbon oxides, nitrogen oxides, VOCs, etc. In order to manage pollution levels multiple sideways were introduced to distribute traffic which results in the distribution of pollutants.

3 Results and Discussions

The results of traffic congestion problem-solving research can vary depending on the specific interventions implemented and the context in which they are implemented. However, the ultimate goal of this research is to reduce traffic congestion and improve the overall efficiency of the transportation system in urban areas.

Certain attributes were observed from both the models, addressing the opportunity for improvement which was identified in the analysis. Observation includes the drop in occupancy rate of the main roads, and uninterrupted flow of traffic with the implementation of traffic signals as shown in Fig. 3.

As traffic signals were introduced there was an increase in queue length in the proposed solution but not lead to traffic congestion as given in Fig. 4. Reduction in queue stops was observed as the majority of the traffic flow was directed toward side roads as shown in Fig. 5.

Pollution level control was one of the unique aspects in this study, which showed acceptable results to improve environmental aspects and to make it fresher at particular node or region. Emissions of COx, NOx, and VOCs were recorded and predicted from the model Figs. 6, 7, and 8 demonstrate respective emissions.

One of the potential results of traffic congestion problem-solving research is a reduction in travel times for commuters. This can be achieved through interventions

Fig. 3 Change in occupancy

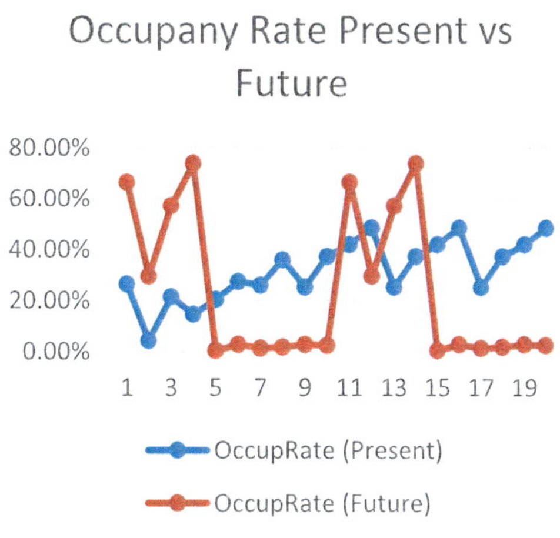

Fig. 4 Change in queue length

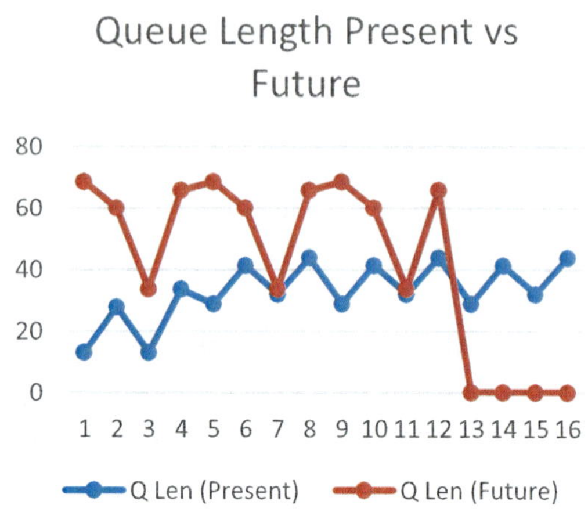

Fig. 5 Reduction in queue stops

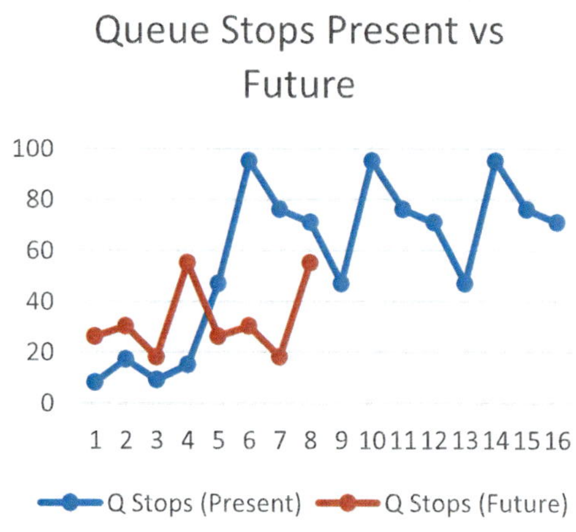

Fig. 6 COx emissions

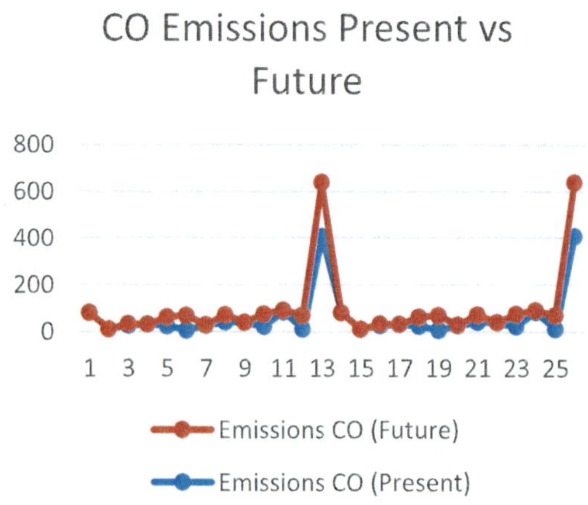

Fig. 7 NOx emissions

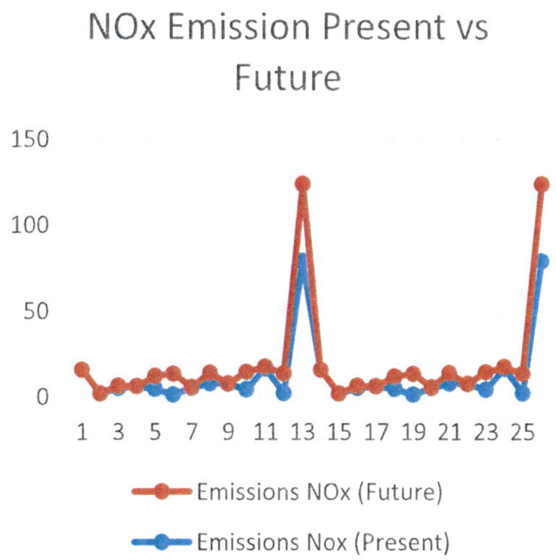

Fig. 8 VOC emissions

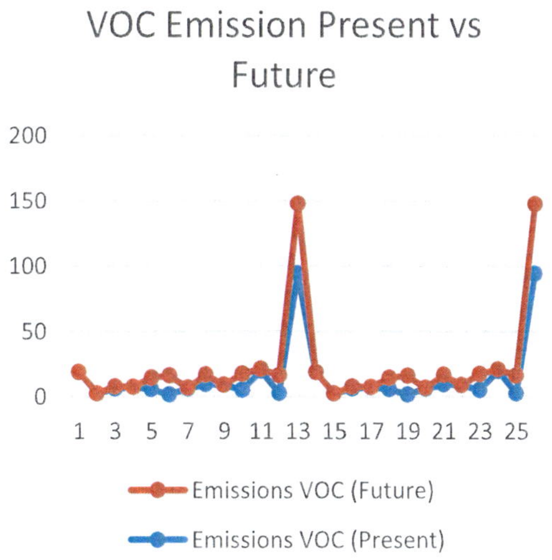

such as adding new lanes to highways, improving public transit systems, or implementing congestion pricing schemes that incentivize drivers to use alternative modes of transportation. In this research, the intervention was made in order to increase the utilization factor of side roads without investing capital to construct new lanes or flyovers.

Conservation of non-renewable energy is another aspect, which uplifts the environment which can be done by the implementation of carpooling as well as due to

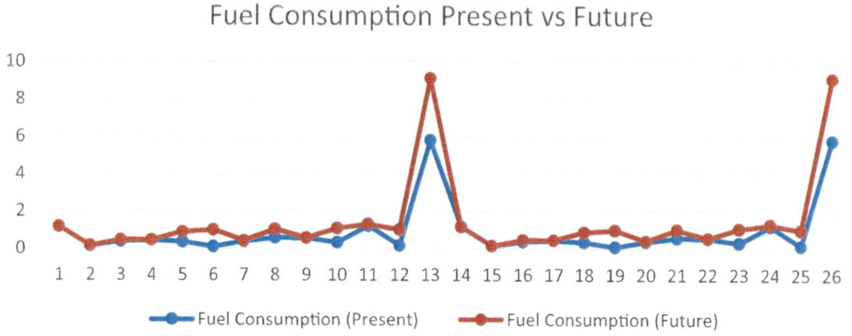

Fig. 9 Change in fuel consumption

the reduction in traffic congestion. Figure 9 shows how the change in fuel consumption takes place as there was a vast reduction in congestion and fuel wastage was minimized.

Another potential result is a reduction in air pollution caused by traffic congestion. By reducing the number of cars on the road or encouraging the use of electric or hybrid vehicles, researchers can help to improve air quality in urban areas and reduce the negative health impacts associated with air pollution. Pollution levels were drastically impacted based on simulation as vehicles were directed to different paths and it was proposed that pollutants at the particular node of road networks will decrease over the period of time.

In addition to these specific outcomes, traffic congestion problem-solving research can also lead to broader discussions and debates about urban planning, transportation policy, and sustainability. By bringing together stakeholders from various fields, this research can help to identify new opportunities for collaboration and innovation, and provide insights into how cities can be designed and managed more effectively to improve the quality of life for residents.

The discussion of these results may involve exploring the challenges and limitations of the interventions implemented and potential trade-offs between different goals. For example, while increasing the use of side roads may reduce travel times and congestions between two nodes, it may also encourage to use of less fuel resulting in energy conservation. As such, researchers may discuss the need for a balanced approach that considers the impacts of interventions across multiple dimensions.

4 Conclusions

Traffic congestion is a big issue in urban areas worldwide. To solve it, a multidisciplinary approach is needed, considering factors such as transportation infrastructure, urban planning, and behavior change. By collecting data, creating models, and developing and evaluating interventions, researchers can find solutions to reduce congestion and improve transportation efficiency. These interventions could include improving public transit, implementing congestion pricing schemes, and promoting alternative transportation modes. Solving traffic congestion is important for sustainable and efficient transportation systems, benefiting residents, businesses, and the environment.

References

1. Abdurakhmanov R (2022) Traffic determination of traffic congestion and delay of traffic flow at controlled intersections. Am J Eng Technol 04:4–11
2. Ziemska M (2021) Exhaust emissions and fuel consumption analysis on the example of an increasing number of HGVs in the Port City. Sustainability 13:14
3. Arnott R, de Palma A, Lindsey R (1989) Does providing information to drivers improve road network reliability?
4. Bando M, Hasebe K, Nakayama A, Shibata A, Sugiyama Y (1995) Dynamical model of traffic congestion and numerical simulation. Phys Rev E 51:1035–1042
5. Arnott R, Small K (1994) The economics of traffic congestion. Am Sci 82:446–455
6. Barth A, Boriboonsomsin M (2008) Real-world CO_2 impacts of traffic congestion. Access Mag 1:24
7. Currie J, Walker R (2011) Traffic congestion and infant health: evidence from E-ZPass. Am Econ J Appl Econ 3:65–90
8. Alghamdi T, Elgazzar K, Bayoumi M, Sharaf T, Shah S (2019) Forecasting traffic congestion using ARIMA modeling. In: 2019 15th international wireless communications and mobile computing conference, IWCMC 2019, pp 1227–1232
9. Harrou F, Zeroual A, Sun Y (2020) Traffic congestion monitoring using an improved kNN strategy. Meas J Int Meas Confed 156:25
10. Wei X, Ren Y, Shen L, Shu T (2022) Exploring the spatiotemporal pattern of traffic congestion performance of large cities in China: a real-time data based investigation. Environ Impact Assess Rev 95:106808
11. Bawaneh M, Simon V (2023) Novel traffic congestion detection algorithms for smart city applications. Concurr Comput Pract Exp 35:1–23
12. Mohammed Almatar K (2023) Traffic congestion patterns in the urban road network: (Dammam metropolitan area). Ain Shams Eng J 14:101886